Building the Arkansas Innovation Economy

Summary of a Symposium

Charles W. Wessner, Rapporteur

Committee on Competing in the 21st Century:
Best Practice in State and Regional Innovation Initiatives

Board on Science, Technology, and Economic Policy

Policy and Global Affairs

NATIONAL RESEARCH COUNCIL
OF THE NATIONAL ACADEMIES

THE NATIONAL ACADEMIES PRESS
Washington, D.C.
www.nap.edu

THE NATIONAL ACADEMIES PRESS   500 Fifth Street NW   Washington, DC 20001

NOTICE: The project that is the subject of this report was approved by the Governing Board of the National Research Council, whose members are drawn from the councils of the National Academy of Sciences, the National Academy of Engineering, and the Institute of Medicine. The members of the committee responsible for the report were chosen for their special competences and with regard for appropriate balance.

This study was supported by: Contract/Grant No. DE-DT0000236, TO# 28, (base award DE-AM01-04PI45013), between the National Academy of Sciences and the Department of Energy; and Contract/Grant No. N01-OD-4-2139, TO# 250, between the National Academy of Sciences and the National Institutes of Health. This report was prepared by the National Academy of Sciences under award number SB134106Z0011, TO# 4 (68059) from the U.S. Department of Commerce, National Institute of Standards and Technology (NIST). This report was prepared by the National Academy of Sciences under award number 99-06-07543-02 from the Economic Development Administration, U.S. Department of Commerce. The statements, findings, conclusions, and recommendations are those of the author(s) and do not necessarily reflect the views of the National Institute of Standards and Technology, the Economic Development Administration, or the U.S. Department of Commerce. Additional support was provided by the Heinz Endowments, the Association of University Research Parks, Acciona Energy, Dow Corning, IBM, and SkyFuel, Inc.

International Standard Book Number 13: 978-0-309-26643-7 (Book)
International Standard Book Number 10: 0-309-26643-2 (Book)

Additional copies of this report are available for sale from the National Academies Press, 500 Fifth Street, NW, Keck 360, Washington, DC 20001; (800) 624-6242 or (202) 334-3313; http://www.nap.edu/ .

Copyright 2012 by the National Academy of Sciences. All rights reserved.

Printed in the United States of America

THE NATIONAL ACADEMIES
Advisers to the Nation on Science, Engineering, and Medicine

The **National Academy of Sciences** is a private, nonprofit, self-perpetuating society of distinguished scholars engaged in scientific and engineering research, dedicated to the furtherance of science and technology and to their use for the general welfare. Upon the authority of the charter granted to it by the Congress in 1863, the Academy has a mandate that requires it to advise the federal government on scientific and technical matters. Dr. Ralph J. Cicerone is president of the National Academy of Sciences.

The **National Academy of Engineering** was established in 1964, under the charter of the National Academy of Sciences, as a parallel organization of outstanding engineers. It is autonomous in its administration and in the selection of its members, sharing with the National Academy of Sciences the responsibility for advising the federal government. The National Academy of Engineering also sponsors engineering programs aimed at meeting national needs, encourages education and research, and recognizes the superior achievements of engineers. Dr. Charles M. Vest is president of the National Academy of Engineering.

The **Institute of Medicine** was established in 1970 by the National Academy of Sciences to secure the services of eminent members of appropriate professions in the examination of policy matters pertaining to the health of the public. The Institute acts under the responsibility given to the National Academy of Sciences by its congressional charter to be an adviser to the federal government and, upon its own initiative, to identify issues of medical care, research, and education. Dr. Harvey V. Fineberg is president of the Institute of Medicine.

The **National Research Council** was organized by the National Academy of Sciences in 1916 to associate the broad community of science and technology with the Academy's purposes of furthering knowledge and advising the federal government. Functioning in accordance with general policies determined by the Academy, the Council has become the principal operating agency of both the National Academy of Sciences and the National Academy of Engineering in providing services to the government, the public, and the scientific and engineering communities. The Council is administered jointly by both Academies and the Institute of Medicine. Dr. Ralph J. Cicerone and Dr. Charles M. Vest are chair and vice chair, respectively, of the National Research Council.

www.national-academies.org

Committee on Competing in the 21st Century:
Best Practice in State and Regional Innovation Initiatives

Mary L. Good, *Chair*
Dean Emeritus, Donaghey College
  of Engineering and Information
  Technology
Special Advisor to the Chancellor
  for Economic Development
University of Arkansas
  at Little Rock
and STEP Board

William C. Harris
President and CEO
Science Foundation Arizona

W. Clark McFadden II
Senior Counsel
Orrick, Herrington & Sutcliffe
LLP.

Michael G. Borrus, *Vice Chair*
Founding General Partner
X/Seed Capital Management

David T. Morgenthaler
Founding Partner
Morgenthaler Ventures

Edward E. Penhoet
Director
Alta Partners

Tyrone C. Taylor
President
Capitol Advisors on Technology

PROJECT STAFF

Charles W. Wessner
Study Director

Alan H. Anderson
Consultant

McAlister T. Clabaugh
Program Officer

David S. Dawson
Senior Program Assistant
(from June 2010)

Sujai J. Shivakumar
Senior Program Officer

David E. Dierksheide
Program Officer

Adam H. Gertz
Program Associate
(through June 2010)

Thomas R. Howell
Consultant

For the National Research Council (NRC), this project was overseen by the Board on Science, Technology and Economic Policy (STEP), a standing board of the NRC established by the National Academies of Sciences and Engineering and the Institute of Medicine in 1991. The mandate of the Board on Science, Technology, and Economic Policy is to advise federal, state, and local governments and inform the public about economic and related public policies to promote the creation, diffusion, and application of new scientific and technical knowledge to enhance the productivity and competitiveness of the U.S. economy and foster economic prosperity for all Americans. The STEP Board and its committees marshal research and the expertise of scholars, industrial managers, investors, and former public officials in a wide range of policy areas that affect the speed and direction of scientific and technological change and their contributions to the growth of the U.S. and global economies. Results are communicated through reports, conferences, workshops, briefings, and electronic media subject to the procedures of the National Academies to ensure their authoritativeness, independence, and objectivity. The members of the STEP Board* and the NRC staff are listed below:

Paul L. Joskow, *Chair*
President
Alfred P. Sloan Foundation

Ernst R. Berndt
Louis E. Seley Professor
 in Applied Economics
Massachusetts Institute
 of Technology

John Donovan
Chief Technology Officer
AT&T Inc.

Alan M. Garber (IOM)
Provost
Harvard University

Ralph E. Gomory (NAS/NAE)
Research Professor
Stern School of Business
New York University

Mary L. Good (NAE)
Dean Emeritus, Donaghey College
 of Engineering and Information
 Technology
Special Advisor to the Chancellor
 for Economic Development
University of Arkansas
 at Little Rock

William H. Janeway
Partner
Warburg Pincus, LLC

Richard K. Lester
Japan Steel Industry Professor
Head, Nuclear Science
 and Engineering
Founding Director, Industrial
 Performance Center
Massachusetts Institute
 of Technology

*As of November 2012.

continued

William F. Meehan III
Lecturer in Strategic Management
Raccoon Partners Lecturer
  in Management
Graduate School of Business
Stanford University
and
Director Emeritus
McKinsey and Co., Inc.

David T. Morgenthaler
Founding Partner
Morgenthaler Ventures

Luis M. Proenza
President
The University of Akron

William J. Raduchel
Chairman
Opera Software ASA

Kathryn L. Shaw
Ernest C. Arbuckle Professor
  of Economics
Graduate School of Business
Stanford University

Laura D'Andrea Tyson
S.K. and Angela Chan Professor
  of Global Management
Haas School of Business
University of California, Berkeley

Harold R. Varian
Chief Economist
Google, Inc.

Alan Wm. Wolff
Senior Counsel
McKenna Long & Aldridge LLP

STEP Staff

Stephen A. Merrill
Executive Director

Paul T. Beaton
Program Officer

McAlister T. Clabaugh
Program Officer

Aqila A. Coulthurst
Program Coordinator

Charles W. Wessner
Program Director

David S. Dawson
Senior Program Assistant

David E. Dierksheide
Program Officer

Sujai J. Shivakumar
Senior Program Officer

Contents

PREFACE   xiii

I. OVERVIEW   1

II. PROCEEDINGS   47

DAY 1

Session I: The Global Challenge and the Opportunity for Arkansas
Moderator: Mary Good, University of Arkansas at Little Rock   49

    The Innovation Imperative: Global Best Practices   49
    Charles Wessner, The National Academies

    Innovation Infrastructure at the State and Regional Level: Some Success Stories   56
    Richard Bendis, Innovation America

    Innovation and Commercialization Successes in Oklahoma   63
    David Thomison, Innovation to Enterprise (i2E)

    California's Innovation Challenges and Opportunities   65
    Susan Hackwood, California Council on Science and Technology

    Evolution of Innovation in Arkansas   70
    Watt Gregory, Accelerate Arkansas

Session II: Cluster Opportunities for Arkansas
Moderator: Paul Suskie, Arkansas Public Service Commission   75

    Arkansas and the New Energy Economy   75
    Paul Suskie, Arkansas Public Service Commission

    Federal-State Synergies   78
    Gilbert Sperling, Energy Efficiency and Renewable Energy (EERE), U.S. Department of Energy

The Wind Energy Industry in Arkansas: **An Innovation Ecosystem** *Joe Brenner, Nordex USA*	85
Arkansas's Role in Energy Transmission Management *Nick Brown, Southwest Power Pool*	87

DAY 2

The State of Technology and Innovation in Arkansas *The Honorable Mike Beebe, Governor of Arkansas*	90
Session II: Cluster Opportunities for Arkansas (continued) *Moderator: Charles Wessner, The National Academies*	93
Research in Advanced Power Electronics: Status and Vision *Alan Mantooth, National Center for Reliable Electric Power Transmission (NCREPT), University of Arkansas at Fayetteville*	93
Regional Initiation Clusters (RIC) *Ginger Lew, National Economic Council, The White House*	96
Agriculture and Food Processing *Carole Cramer, Arkansas Biosciences Institute, Arkansas State University*	99
Information Technology *Jeff Johnson, ClearPointe*	101
Nanotechnology *Greg Salamo and Alex Biris, University of Arkansas at Fayetteville*	104
Session III: Federal and State Programs and Synergies *Department of Commerce*	107
The Role of the Economic Development Administration *Barry Johnson, Economic Development Administration, Department of Commerce*	107
Initiatives of the Manufacturing Extension Program *Roger Kilmer, Manufacturing Extension Partnership, National Institute of Standards and Technology*	110

University-Industry Partnerships — 113
Marc Stanley, National Institute of Standards and Technology

University-Federal Government Partnerships — 119
Donald Senich, Division of Industrial Innovation and Partnerships, Directorate of Engineering, National Science Foundation

From University Research to Start-ups: Building Deals for Arkansas — 121
Michael Douglas, UAMS BioVentures, University of Arkansas Medical Services

Session IV: Universities and Regional Growth
Moderator: John Ahlen, Arkansas Science and Technology Authority — 125

Arkansas STEM Coalition Activities — 125
Michael A. Gealt, University of Arkansas at Little Rock

State Initiatives for Research Funding and Their Role in Economic Development — 128
William Harris, Science Foundation Arizona

Session V: Arkansas R&D Capacity: Universities, Research Labs, and Science Parks
Moderator: John Ahlen, Arkansas Science and Technology Authority — 134

Infrastructure for High-Performance Computing — 134
Amy Apon, High-Performance Computer Center, University of Arkansas at Fayetteville, and Division of Computer Science, Clemson University

Research Parks in Arkansas — 136
Jay Chesshir, Little Rock Chamber of Commerce

Understanding the Battelle Study — 138
Jerry Adams, Arkansas Research Alliance

III. APPENDIXES

	A	Agenda	143
	B	Participants List	147
	C	Bibliography	151

Preface

Responding to the challenges of fostering regional growth and employment in an increasingly competitive global economy, many U.S. states and regions have developed programs to attract and grow companies as well as attract the talent and resources necessary to develop innovation clusters. These state and regionally based initiatives have a broad range of goals and increasingly include significant resources, often with a sectoral focus and often in partnership with foundations and universities. These are being joined by recent initiatives to coordinate and concentrate investments from a variety of federal agencies that provide significant resources to develop regional centers of innovation, business incubators, and other strategies to encourage entrepreneurship and high-tech development.

PROJECT STATEMENT OF TASK

An ad hoc committee, under the auspices of the Board on Science, Technology, and Economic Policy (STEP), is conducting a study of selected state and regional programs in order to identify best practices with regard to their goals, structures, instruments, modes of operation, synergies across private and public programs, funding mechanisms and levels, and evaluation efforts. The committee is reviewing selected state and regional efforts to capitalize on federal and state investments in areas of critical national needs. This review includes both efforts to strengthen existing industries as well as specific new technology focus areas such as nanotechnology, stem cells, and energy in order to better understand program goals, challenges, and accomplishments.

As a part of this review, the committee is convening a series of public workshops and symposia involving responsible local, state, and federal officials and other stakeholders. These meetings and symposia will enable an exchange of views, information, experience, and analysis to identify best practice in the range of programs and incentives adopted.

Drawing from discussions at these symposia, fact-finding meetings, and commissioned analyses of existing state and regional programs and technology focus areas, the committee will subsequently produce a final report with findings and recommendations focused on lessons, issues, and opportunities for complementary U.S. policies created by these state and regional initiatives.

THE CONTEXT OF THIS PROJECT

Since 1991, the National Research Council, under the auspices of the Board on Science, Technology, and Economic Policy, has undertaken a program of activities to improve policymakers' understandings of the interconnections of science, technology, and economic policy and their importance for the American economy and its international competitive position. The Board's activities have corresponded with increased policy recognition of the importance of knowledge and technology to economic growth.

One important element of STEP's analysis concerns the growth and impact of foreign technology programs.[1] U.S. competitors have launched substantial programs to support new technologies, small firm development, and consortia among large and small firms to strengthen national and regional positions in strategic sectors. Some governments overseas have chosen to provide public support to innovation to overcome the market imperfections apparent in their national innovation systems.[2] They believe that the rising costs and risks associated with new potentially high-payoff technologies, and the growing global dispersal of technical expertise, underscore the need for national R&D programs to support new and existing high-technology firms within their borders.

Similarly, many state and local governments and regional entities in the United States are undertaking a variety of initiatives to enhance local economic development and employment through investment programs designed to attract knowledge-based industries and grow innovation clusters.[3] These state and regional programs and associated policy measures are of great interest for their potential contributions to growth and U.S. competitiveness and for the "best practice" lessons they offer for other state and regional programs.

STEP's project on State and Regional Innovation Initiatives is intended to generate a better understanding of the challenges associated with the transition of research into products, the practices associated with successful state and regional programs, and their interaction with federal programs and private initiatives. The study seeks to achieve this goal through a series of complementary assessments of state, regional, and federal initiatives; analyses of specific industries and technologies from the perspective of crafting supportive public policy at all three levels; and outreach to multiple

---

[1] National Research Council, *Innovation Policies for the 21st Century, Report of a Symposium*, C. Wessner, ed., Washington, DC: National Academies Press, 2007.
[2] For example, a number of countries are investing significant funds in the development of research parks. For a review of selected national efforts, see National Research Council, *Understanding Research, Science and Technology Parks: Global Best Practices, Report of a Symposium*, C. Wessner, ed., Washington, DC: National Academies Press, 2009.
[3] For a scoreboard of state efforts, see Robert Atkinson and Scott Andes, *The 2010 State New Economy Index: Benchmarking Economic Transformation in the States,* Kauffman Foundation and ITIF, November 2010.

PREFACE

stakeholders. The overall goal is to improve the operation of state and regional programs and, collectively, enhance their impact.

THIS SUMMARY

The symposium reported in this volume convened state officials and staff, business leaders, and leading national figures in early-stage finance, technology, engineering, education, and state and federal policies to review challenges, plans, and opportunities for innovation-led growth in Arkansas. The symposium included an assessment of Arkansas' natural, industrial, and human resources; an identification of key sectors and issues; and a discussion of how the state might leverage national programs to support its economic development goals.

This summary includes an introduction that highlights key issues raised at the meeting and a summary of the meeting's presentations. This workshop summary has been prepared by the workshop rapporteur as a factual summary of what occurred at the workshop. The planning committee's role was limited to planning and convening the workshop. The statements made are those of the rapporteur or individual workshop participants and do not necessarily represent the views of all workshop participants, the planning committee, or the National Academies.

ACKNOWLEDGMENTS

On behalf of the National Academies, we express our appreciation and recognition for the insights, experiences, and perspectives made available by the participants of this meeting. We are indebted to Alan Anderson for summarizing the proceedings of the meeting and to Tom Howell for preparing the draft introduction. We are also indebted to Sujai Shivakumar and David Dawson of the STEP staff for preparing the report manuscript for publication.

NATIONAL RESEARCH COUNCIL REVIEW

This report has been reviewed in draft form by individuals chosen for their diverse perspectives and technical expertise, in accordance with procedures approved by the National Academies' Report Review Committee. The purpose of this independent review is to provide candid and critical comments that will assist the institution in making its published report as sound as possible and to ensure that the report meets institutional standards for quality and objectivity. The review comments and draft manuscript remain confidential to protect the integrity of the process.

We wish to thank the following individuals for their review of this report: John Ahlen, Arkansas Science & Technology Authority; Edward Malecki, Ohio State University; Lora Lee Martin, California Council on Science and Technology; and Eric Sandgren, University of Arkansas, Little Rock.

Although the reviewers listed above have provided many constructive comments and suggestions, they were not asked to endorse the content of the report, nor did they see the final draft before its release. Responsibility for the final content of this report rests entirely with the rapporteur and the institution.

Charles W. Wessner
Mary L. Good

I

OVERVIEW

Overview

Arkansas is seeking to reinvent itself as a knowledge-based economy. This transformation continues the state's longstanding efforts to adapt to changing economic conditions. When the post-World War II automation of agriculture displaced much of the state's predominantly agrarian work force, the state succeeded in attracting manufacturing industries based on Arkansas' low wages and favorable business climate. When these industries began moving offshore in the 1970s, the state experienced a steady erosion of manufacturing jobs that continues to the present day. Arkansas began building the infrastructure for technology-based economic development in the 1980s. Recognizing the growing importance of knowledge-based industries, Arkansas' leaders are once again taking steps to secure their future economic growth.

To review the states recent initiatives to develop a knowledge-based economy, the National Academies' Board on Science, Technology, and Economic Policy (STEP) and University of Arkansas at Little Rock convened a conference on *Building the Arkansas Innovation Economy.* Held at the Clinton Presidential Library in Little Rock, the conference drew Arkansas business, political, and academic leaders along with senior U.S. government officials and national experts to highlight the accomplishments and growth of the innovation ecosystem in Arkansas, while also identifying needs, challenges, and opportunities. The participants at this conference discussed a series of proposed initiatives to strengthen Arkansas' innovation and technology infrastructure and identified areas where federal, state, and foundation contributions could generate positive synergies.

As this report of the conference documents, Arkansas' business, academic and government leaders recognize the economic and technological challenges confronting the state. They have studied successful economic and research programs in other states and drawn on national experts to develop strategic plans to promote economic growth and – in recent years – to enhance the state's standing in the knowledge-based economy of the 21st century. Arkansas benefits from homegrown entrepreneurial ingenuity and pluck, its reputation as a highly pro-business state, strong transportation links, and a geographic location in the center of the North American market.

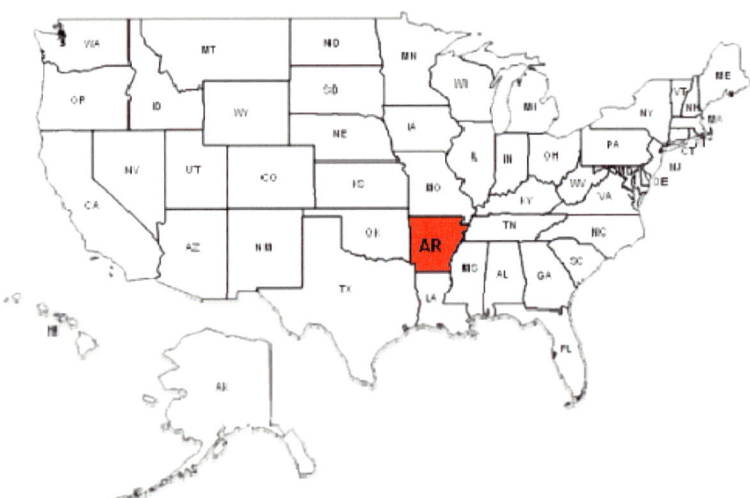

FIGURE 1 Location of Arkansas in the United States

At the same time, the state's development continues to be hampered by weaknesses in its knowledge and skills base, the out-migration of college graduates, a dearth of venture capital, and a relatively low level of federal research funding (See Table 1). Many of the Arkansas' economic and technology development initiatives were designed to address these areas of vulnerability.

Table 1 Innovation Indicators for Arkansas		
Category	Arkansas	U.S. Average
State funding for major public universities per enrolled student (2010)	$10, 825	$9,082
Engineers in the workforce (2010)	0.53	1.12
Life and physical scientists in the workforce (2010)	0.3	0.45
Federal R&D obligation per employed worker (2008)	$112	$862
Federal R&D obligations per S&E occupation holder (2008):	$4,947	$21,594
Academic S&E articles per 1000 science doctorate holders in academia (2008)	477	579
Patents per 1000 S&E occupation holders (2010):	4.9	19.4
Venture Capital deals per high tech establishments (2008):	0.0	0.59
High tech to all business establishments (2008)	7.02	8.52
Business performed R&D to private industry output (2008)	0.52	2.14

Source: NSB Science and Engineering Indicators 2012 State Data Tool

THE INNOVATION IMPERATIVE

The twenty-first century is witness to fundamental changes in the world's economies. Knowledge-based economic activity is recognized worldwide as the basis for sustained growth. The prosperity of individual regions is based increasingly on their relative success in attracting and retaining knowledge-based activities and assets and utilizing them for economic development. At the same time, the globalization of trade and investment as well as advances in communications and transportation has created an increasingly integrated global market. Reflecting the growing mobility of capital and labor, states and regions are increasingly vulnerable to companies, industries and jobs moving to other parts of the U.S. or to foreign countries that offer a skilled and flexible workforce, often at lower cost, and greater incentives for investment. The nation's states and regions therefore face an imperative to foster innovation and start, grow, and retain innovative firms if they are to sustain and augment their standard of living and ensure their long run economic well-being.[1]

BOX 1:
Addressing the Innovation Imperative in Arkansas

In his conference keynote, Governor Beebe of Arkansas said that among the state's advantages are its strong work ethic and entrepreneurial spirit. He noted Arkansas' reputation for successful businesses, beginning with Sam Walton's Wal-Mart, and continuing with Tyson, J. B. Hunt, Stevens, Acxiom, and Alltel, which had become part of Verizon. "Those success stories were the basis for what was yesterday," he said, "but they provide us with a roadmap for tomorrow. We're probably not even aware of how our children and grandchildren will live 10, 15, or 20 years from now. But those who embrace technology and innovation and entrepreneurship; make the marriage between education and economic development; and learn that science is the basis for tomorrow's economy will reap the benefits for themselves, their employees, their loved ones, and their region."

Acknowledging that the state lags in per capita baccalaureate degrees, where it stands 49th in the nation, Governor Beebe said that he was determined to change that ranking.[2] The state has initiated policies that include higher standards, higher expectations, and more advanced placement. The state has approved a lottery, with all of its available revenues targeted for college scholarships. "There will be no excuse for Arkansas to stay 49th in per capita BA degrees," he said.

In his conference remarks, Richard Bendis, of Innovation America, outlined the key issues for building knowledge and innovation-based economies. He defined innovation as "the creation and transformation of knowledge into new products, processes and services that meet market needs." Knowledge economies are "based on creating, evaluating and trading knowledge." Bendis observed that the public, academic and private sectors each have essential roles in innovation. Academia must focus on the creation, integration and transfer of

---

[1] Ross De Vol, et al., *Arkansas' Position in the Knowledge-Based Economy* Santa Monica: Milken Institute, September 2004, p. 1.
[2] Recognizing these realities, *Governor Mike Beebe's Strategic Plan for Economic Development* points out that Arkansas is "at a critical disadvantage in competing for opportunities in the 21st century economy," and that the state had "not kept pace" with the requirements of the global knowledge-based economy. See, the Executive Summary of *Governor Mike Beebe's Strategic Plan for Economic Development, Little Rock:* Arkansas Economic Development Commission, 2009.

knowledge itself. Industry must use that knowledge to create wealth in the form of commercial products, processes and services. Governments must develop policies to encourage innovation, engage in long-term vision and planning, invest in under-supported areas, and participate in public-private partnerships with industry.[3]

He added that technology clusters, which facilitate innovation through physical proximity and the close interaction of many actors from different sectors, serve as catalysts for innovation and the creation of start-ups. While Silicon Valley, Route 128 in Boston and Research Triangle Park in North Carolina are widely cited as examples of successful technology clusters, growing successful clusters by stimulating the development, commercialization and financing of technology-based firms is a significant challenge.[4] In this regard, Bendis stressed the role of "innovation intermediaries" to coordinate local technologies, assets and resources to advance innovation in a jurisdiction.[5]

Bendis further observed that successful development of economic activity within a region requires a "three-legged stool." First, the region must attract companies from other regions; second, it must retain companies that are already present; and third, it must create new companies. Creating new companies is both the most important element and the one that is most difficult to achieve. It is important because small and medium enterprises (SMEs) are a major source of innovation and new jobs.[6] However, as Jeff Johnson, CEO and President of ClearPointe Technology, a managed service provider based in Little Rock, noted at the conference, "very few new firms have adequate cash to get a new business through the first year, and we were no exception."[7] The result, according to Bendis, is that most start-ups with new ideas do not move to the commercialization stage – at present of 150-200 small firms that develop business plans, only about 10 draw the interest of venture capitalists, and only one is actually funded. Most small firms that need financial backing are in the proof-of-concept, start-up or seed capital phases, and typically need $500,000 to $2 million for the development of prototypes.[8] This need is not being met; seed

---

[3] See the summary of the presentation by Richard Bendis in the Proceedings chapter in this volume.
[4] For a review of the nature of innovation clusters and state policies to grow innovation clusters, see National Research Council, *Growing Innovation Clusters for American Prosperity,* Charles W. Wessner, rapporteur, Washington, D.C.: The National Academies Press, 2011.
[5] See the summary of the presentation by Richard Bendis in the Proceedings chapter in this volume.
[6] Small and medium enterprises generate 13 times as many patented technologies as large firms, and are an important source of innovation for large firms that often partner with SMEs. In the three years after the recession of 2001, companies of less than 20 employees created 107 percent of net new jobs while companies over 500 employees eliminated a net of -24 percent. See the summary of the presentation by Richard Bendis in the Proceedings chapter in this volume, where he cites data from the Small Business Administration.
[7] See the summary of the presentation by Jeff Johnson in the Proceedings chapter in this volume.
[8] See the summary of the presentation by Richard Bendis in the Proceedings chapter in this volume.

stage investments by the U.S. venture capital industry declined by 48 percent in 2011 to $919 million, or 3 percent of all venture capital investment.[9] In effect, venture investment is moving downstream, away from risk, a phenomenon that is acting as a drag on start-ups in Arkansas and elsewhere.

Responding to this innovation imperative, governments around the world have implemented a variety of policies and programs designed to promote innovation-based economic growth.[10] Many of these efforts emulate U.S. public-private initiatives that are widely seen as successful. Indeed, the United States has a long tradition in public-private partnerships, beginning with a 1798 government grant to Eli Whitney to produce muskets with interchangeable parts, and continuing through government support for development of the telegraph, the airplane, jet aircraft, semiconductors, computers, nuclear energy and satellites.[11]

As we see next, Arkansas' recent economic and technological development efforts are a part of this long national tradition in cooperation and pragmatism in fostering economic growth and addressing common missions.

BACKGROUND: ECONOMIC DEVELOPMENT EFFORTS IN ARKANSAS, 1955-2012

For most of its history, farming and forestry dominated Arkansas' economy. During the mid-20th century, sharecroppers and poor migrant laborers were displaced by agricultural automation. While many migrated, others stayed, making up the state's pool of low cost surplus labor. In 1950, Arkansas' per capita income was 56 percent of the national average, and its population was declining. In 1955, the state legislature established the Arkansas Industrial Development Commission (AIDC) with a mandate to promote industrial development. Under the leadership of its first chairman, Winthrop Rockefeller, the AIDC began to court out-of-state businesses aggressively.[12] The result was

---

[9] PWC MoneyTree Venture Capital Report, 2010.
[10] For a review of innovation polices of leading nations and regions in Asia, Europe, and North America and the challenges facing the United States, see National Research Council, *Rising to the Challenge, U.S. Innovation Policy for the Global Economy,* C. Wessner and A., Wm., Wolff, eds., Washington, DC: National Academies Press, 2012.
[11] For an abridged history of US public private partnerships and an analysis of factors characterizing successful partnerships, see National Research Council, *Government Industry Partnerships for the Development of New Technologies*, C. Wessner, ed., Washington, DC: National Academies Press, 2003.
[12] The AIDC was formed pursuant to Act 404 of 1955, which also authorized incorporation of local industrial development corporations and issuance of local industrial development bonds. Today the AIDC operates as the Arkansas Economic Development Commission (AEDC), with a current mandate to promote economic development and develop strategies that produce better-paying jobs, support communities and support workforce training. See the summary of the presentation by Watt

an influx of manufacturers seeking low wage labor and cheap land. Between 1955 and 1960 Arkansas added over 51,000 jobs.[13] In 1997, AIDC was renamed the Arkansas Economic Development Commission (AEDC) to reflect a broader emphasis on developing service and high technology industries in the state.

By the 1970s the "smokestack" industries that had located in Arkansas during the preceding decades came under international competitive pressure and began to move offshore. The percentage of manufacturing employment in the state began a long decline, from 32 percent in 1975 to 17 percent by 2005. Per capita income rose through the 1960s and 1970s, but leveled off in 1978 at about 77 percent of the national average where it has "refused to budge despite the best efforts of economic developers in the state."[14] Arkansas' unemployment rate, which stood at 5.0 percent in 1970, nearly doubled to 9.5 percent in 1975. Unemployment peaked at 9.7 percent in 1983 and remained above seven percent through the entire decade of the 1980s.[15] In 1979, the AEDC released a report that warned that the state's future economic growth was limited by a strategy that sought to recruit firms that provided labor-intensive, low skill, minimum wage jobs to Arkansans.[16]

During the 1980s the state took a number of steps to counteract the loss of businesses and jobs. Two new development agencies were established: The Arkansas Science & Technology Authority (ASTA) was tasked with promoting innovation, scientific research, and science, technology, engineering and mathematics (STEM) education.[17] The Arkansas Development Finance Authority (ADFA) was established to provide tax-exempt bonds to finance businesses and education.[18]

Gregory in the Proceedings chapter in this; See also Governor *Mike Beebe's Strategic Plan for Economic Development,* 2009, op. cit., p. 2.
[13] Governor Mike Beebe's *Strategic Plan* (2009), op. cit., p. 3.
[14] Report of the Accelerate Arkansas Strategic Planning Committee, *Building a Knowledge-Based Economy in Arkansas: Strategic Recommendations by Accelerate Arkansas* (September 2007), pp. 16-17.
[15] Governor Mike Beebe's *Strategic Plan*, (2009), op. cit., p. 5.
[16] AEDC, *Arkansas Climbs the Ladder: A View of Economic Factors Relating to Growth of Jobs and Purchasing Power* (1979).
[17] Galley Support Innovations (GSI) is a California-based manufacturer of galley locks and latches for OEMs in the aerospace business that relocated to Arkansas in 2005. Hit hard by the economic downturn that began in 2008, it sought assistance from the Arkansas Science and Technology Authority's Arkansas Manufacturing Solutions program. GSI was awarded to ASTA Technology Transfer Assistance Grants (TTAG), which enabled GSI to secure a large multi-year contract with an estimated positive financial impact of $5 million over six years. "AMS Grant Helps Local Aerospace Manufacturer Turn Business Around," *Arkansas Business* (January 5, 2012).
[18] (2009) op. cit., p. 6. The Arkansas Science and Technology Authority was created by statute in 1983 to "support scientific and business innovation as an economic development tool." In 2009, it completed 31 projects involving about $8 million in grants and tax credits. It has provided grants to support the Arkansas High-Performance Computing Center at the University of Arkansas and to the

Despite difficult economic circumstances, Arkansas has spawned a significant number of world-class companies.[19] Wal-Mart, which began with one retail outlet in Rogers, Arkansas, in 1962, became the nation's largest retailer in 1991.[20] J.B. Hunt, founded in 1961 in Lowell, Arkansas with five trucks and seven refrigerated trailers, became the largest US trucking company by the early 1990s. Tyson Foods, based in Springdale, Arkansas, and which originally consisted of a farmer driving a single truck to deliver chickens to Chicago, became the largest U.S. processor of poultry and the world's second largest processor of chicken, beef and pork. Murphy Oil Corporation, based in El Dorado, Arkansas, operates onshore and offshore oil and natural gas drilling operations globally, and in 2008 ranked 134th on the Fortune 500 list.[21] Other major companies with origins in the state include Riceland, Stephens Inc., Dillard's, Alltel and Acxiom. While most of these companies are not regarded as technology-intensive firms, many of them have applied technology in their business processes with dramatic and in some cases revolutionary impact.[22]

Arkansas Research and Education Optical Network (ARE-ON), a communications network linking Arkansas' four-year public universities. The authority provides financial support for technology transfer to local businesses, provides working capital for small start-up businesses (usually pursuant to royalty-based agreements), and sponsors professional development workshops for teachers and grants to individual STEM teachers for equipment and supplies. See the presentation by Watt Gregory, "Evolution of Innovation in Arkansas," in the Proceedings chapter of this volume. See also the Arkansas Science and Technology Authority website, http://www.asta.arkansas.gov accessed January 11, 2012.

[19] Giang Ho and Anthong Pennington-Cross, "Fayetteville and Hot Springs Lead the Recovery in Employment," *The Regional Economist* (October 2005).

[20] Wal-Mart, with one of the most sophisticated and innovative supply chains in the world, has attracted distribution centers from its major vendors to Arkansas, including Heinz, Clorox, Pfizer, General Mills, Mattel, PepsiCo, Procter & Gamble, Coca-Cola, Johnson & Johnson and Hershey's. "Arkansas: A Natural Wonder," *Inbound Logistics* (May 2009).

[21] The company offers to pay college tuition and fees for all El Dorado high school students. "Murphy Oil Company," *Arkansas Business*
http://www.arkansasbusiness.com/company_info.asp?sym=MUR

[22] Wal-Mart's emergence as the world's largest retailer and the world leader in supply chain logistics is attributed largely to its pioneering practice of tracking inventory by high performance computers.

[22] See the summary of the presentation by Watt Gregory in the Proceedings chapter in this volume.

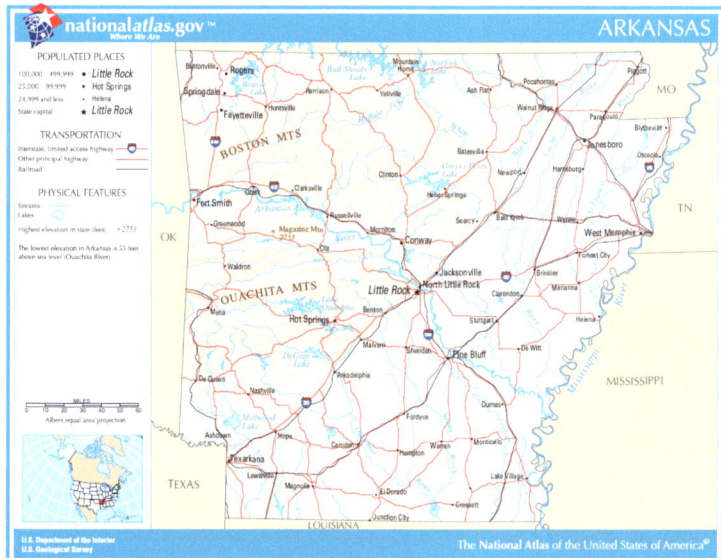

FIGURE 2 Map of the State of Arkansas
Source: U.S. Department of Interior, U.S. Geological Survey

Arkansas' location in the center of the North American market, reinforced by a strong rail, highway and river-based transportation network, (see Figure 2) has proven a major advantage with respect to attracting and holding some traditional industries.[23] Some local manufacturers have found that Arkansas' location lowers transportation costs and thus makes them cost-competitive with products made in China.[24] It is becoming evident that Arkansas enjoys geographic

---

[23] Arkansas has over 1,000 miles of navigable waterways and port facilities on the Mississippi and Arkansas Rivers. Three carriers provide rail service including intermodal freight service. Eight interstate highways cross various parts of the state. Nucor Steel located a mini-mill in Hickman, Arkansas, which makes thin-slab steel coils for use in pipes, tubes, processors and automotive applications, citing location and transportation infrastructure as key decisional factors. Mike Parrish, a Nucor Vice President who managed the Hickman plant, said in 1996 that "What's great about this area is it's centrally located in the country. It's great to advertise in this area. You're on the [Mississippi] River. You can not only ship anywhere in the country, you can ship anywhere in the world," "Nucor Makes Blytheville Steel Capital of the South," *Arkansas Business* (December 16, 1996).

[24] In 2011, the designers of five-gallon Kosmo coolers, a proprietary product, abandoned plans to have them manufactured in China, a decision based on "prohibitive" shipping costs. Instead most of the parts for the coolers will be manufactured by custom injection and blow molder River Bend Industries LLC at its factory in Fort Smith, Arkansas. River Bend worked with the designers, Arkansas entrepreneurs Tim Mika and Steve Bowman, developed a mostly US-made product (only the metal legs will be sourced from China) containing blow-molded and injection molded parts and

advantages with respect to some technology-intensive activities. Companies such as Nordex USA, LM Glasfiber and Mitsubishi Power System Americas, makers of sophisticated wind power equipment, have all cited the state's location as a key factor in their decision to locate major manufacturing facilities there.[25] These traditional advantages, however, may not prove sufficient to address the competitive challenges of the 21st century.

Competing for Industry with Incentives

For decades, state, county and municipal governments across the United States have offered various incentives to attract companies to their jurisdictions. While frequently criticized by economists as inefficient and a misallocation of public resources, jurisdictions that refrain from incentives competition risked the loss of companies, industries and jobs to other areas.[26] Arkansas state, county and municipal governments have repeatedly sought to be competitive with other states and regions by using incentives to attract and retain companies. The state was one of the first in the U.S. to enact tax incentives to attract out-of-state companies.[27] The Arkansas General Assembly enacted special incentives legislation tailored to the needs of Nucor Corp., a highly competitive steelmaker, for a new mill in Blytheville in the 1980s, and the company continued to receive incentives for many years thereafter as it expanded its presence in the state.[28] Arkansas county and municipal

---

urethane from insulation. The Fort Smith location made River Bend's coolers cost-competitive with China. To build the coolers River Bend is expanding its manufacturing capacity in Fort Smith. "River Bend Gets Kosmo Work," *Plastics News* (November 7, 2011).

[25] "LM Glasifiber Dedicates Little Rock Factory," *ArkansasOnline* (October 28, 2008). Senior Vice President Ichiro Fukue commented that "transportation is very important for this industry. Fort Smith is the center of our market." "Arkansas Wins $100 Million Wind Turbine Nacelle Plant," *Energy Overviews* (May 11, 2011). In 2010, Mitsubishi Power System Americas broke ground on a $100 million wind turbine nacelle plant in Fort Smith, Arkansas, citing the location's transportation advantages as a key factor in the site selection. A Nordex executive commented "here in Jonesboro, we're very near the Mississippi River for barging, there are two railways crossing Arkansas, and the highway system provides a major transportation network, so we can ship the 2.5 MW "Gamma Generation" turbines we're building here anywhere in the country and make delivery in a matter of days." Interview with Joe Brenner, Vice President for Production, in *Wind Systems* (January 2011).

[26] Testimony of Arthur J. Rolnick Director of Research, Federal Reserve Bank of Minnesota, before the House Domestic Policy Subcommittee, "Congress Should End the Economic War Among the States," October 10, 2007; Kenneth P. Thomas, *Investment Incentives and the Global Competition for Capital* (London and Basingstoke: Palgrave MacMillan, 2011).

[27] The Enterprise Zone Act of 1983 established tax credits for creation of new jobs and sales and use tax exemptions for building materials used to construct factories. The Manufacturers' Investment Tax Credit (later renamed InvestArk) extends manufacturers sales and use tax credits equal to seven percent of the cost of modernizing and expanding facilities. Governor Mike Beebe's *Strategic Plan* (2009) p. 6.

[28] See Thomas Howell, "Incentives Competition for Businesses," In the Appendix to this report. Between 1996 and 1998, Nucor's Blytheville project reportedly received $48.6 million in tax credits

governments have also achieved significant successes in competing for investment through use of incentive packages.[29] In 2007, the General Assembly approved the creation of the $50 million Quick Action Closing Fund, designed to enable the Governor to "act quickly and decisively in highly competitive situations to finalize an agreement with a company to locate in Arkansas," and the fund has been deployed effectively on a number of occasions.[30] Arkansas continues to make extensive use of incentives – frequently combined packages of state, county and municipal benefits – to compete for investments in emerging sectors, such as wind turbines.[31] Incentives have also played a role in the retention of established knowledge-based companies and jobs.[32] An

---

and cash. "Results from Subsidies Unknown – State Has Little Idea Whether $633 Million in Breaks to Firms Spurred Investment," *Arkansas Democrat-Gazette* December 12, 1999); "Choosing a Greenfield Site: Steelmakers Are Attracted Rural Areas," *Iron Age* (March 1992); "Arkansas Legislators Present Their Proposal for Tax Breaks for Proposed Steel Mill," *Arkansas Democrat-Gazette* (December 7, 1987).

[29] Osceola, Arkansas, was shaken in 2000-01 by the loss of manufacturing plants shut down by Fruit of the Loom (textiles), Eck Adams (furniture), Siegel-Robert (auto parts) and Southwire (wire). The city began using cash generated by the municipally owned electricity distribution system to fund industrial development. In 2003, Osceola secured a new factory opened by DENSO Mfg., a Japanese maker of automotive ventilation systems, defeating other localities with a $3 million combination of incentives, which included land, site improvements, and subsidized electricity rates. Shortly thereafter, Osceola won a $1.2 billion coal fired power plant, the Plum Point Energy Station, offering a 1,000-acre site with infrastructure improvements and a 20-year estate tax abatement. Susan C. Thompson, "Factory Closing Shock Community into Opening Wallets for Economic Development," *The Regional Economist* (October 2010).

[30] Letter from AEDC Executive Director Maria Haley to Senator Mary Anne Salmon and Representative Tommy Lee Baker, Arkansas Legislative Council, August 22, 2011. In 2010, Caterpillar Inc. opened a $140 million, 600-job road grader factory in North Little Rock. Governor Beebe, who used $3 million from the fund to close the deal, said that "They (Caterpillar) wouldn't be here without the Quick Action Closing Fund." "Caterpillar Opens New Arkansas Factory, Hiring 600," *Cleveland.com* (September 1, 2010).

[31] In 2010, Arkansas secured an agreement with Japan's Mitsubishi Power Systems America, Inc. for establishment of a $100 million plant to manufacture wind turbine nacelles at Fort Smith, Arkansas. The city of Fort Smith agreed to issue up to $75 million in revenue bonds for plant construction, to give Mitsubishi a free 90-acre site, to extend streets, water mains and sewers to the plant, and to give Mitsubishi a 50 percent reduction on property taxes. Sebastian County agreed to designate its entire $3.7 million allocation for federal recover zone bank to Mitsubishi. Governor Beebe contributed $3.75 million from the Quick Action Closing Fund to close the deal. "Mitsubishi Incentives Hit $83M," *Fort Smith Times Record* (December 25, 2010).

[32] Windstream, a rapidly-growing telecommunications network created in 2006 in Little Rock, with $5 billion in annual revenues, investigated alternative sites outside the state for its permanent headquarters, but decided in 2010 to stay in Little Rock, with CEO Jeff Gardner acknowledging that "financial incentives provided by the state and the regional placement relative to the company's placement relative to the company's customers in the southeast played into the decision. The incentives included a $1 million in Quick Action Closing Funds for building and training and additional benefits tied to performance. "Windstream Picks Little Rock, AR for HQ," *Business Facilities* (July 13, 2010).

example is the attraction of a Hewlett-Packard Technology Center to Central Arkansas in 2009.

The Growing Importance of Knowledge-based Industries

In 1998, with the state's manufacturing jobs eroding and college graduates out-migrating, then Governor Mike Huckabee convened the Governor's Summit on Economic Development to consider state policies to promote economic growth, development and job creation. The Summit developed several recommendations that were implemented by the state's General Assembly in the form of new legislation. Existing incentive programs were extended to include knowledge-based industries, capital gains tax rates were cut by 30 percent, an Arkansas Research Matching Fund was created, and a small business loan pilot program was established. The Tuition Reimbursement Tax Credit Program was established, authorizing a 30 percent tax credit for targeted companies for costs paid to employees to improve their post-secondary school education.[33]

In 2001, the Director of the Arkansas Department of Economic Development formed the Task Force for the Creation of Knowledge-Based Jobs, comprised of business, academic and government leaders, to develop strategies for expanding the number of knowledge-based jobs and companies in Arkansas. The task force's report, released in September 2002, recommended that math and science be elevated to the number one educational issue in the state; that up to six research centers of excellence be established; that a variety of measures be taken to provide financial support for Arkansas-based technology start-up companies and that the state constitution be amended to permit equity investments by qualified state agencies.[34]

The Milken Study of Arkansas' Competitiveness

Accelerate Arkansas, a statewide group of business leaders working on a volunteer basis to promote knowledge-based institutions, industries, workers, and partnerships, emerged from the Task Force and other state economic development initiatives. Speaking at the conference, Watt Gregory, the chair of Accelerate Arkansas, said that his organization commissioned studies about the Arkansas economy, most importantly a landmark report by the Milken Institute that represented the most comprehensive study ever undertaken of the Arkansas economy.[35] The Milken study observed that a fundamental transformational

---

[33] Governor Mike Beebe's *Strategic Plan* (2009), op. cit., p. 8.
[34] *Report of the Task Force for the Creation of Knowledge-Based Jobs* (September 2002), pp. 2-3.
[35] See the summary of the presentation by Watt Gregory in the Proceedings chapter in this volume.

shift was occurring in the world toward knowledge-based economic activity, reducing the relative importance of traditional assets of regional economic development — proximity to railroads, waterways and raw materials — and factors such as low labor and business costs. Regional economic prosperity, the study concluded, was increasingly based on the relative success of a location in attracting knowledge-based investments and leveraging them for economic development. The study observed that Arkansas had been "operating at the periphery of the knowledge-based economy," — and while the state had been making progress, it was "starting far behind other states in the knowledge-based economy race."[36]

Using the Milken Institute's State Technology and Science Index, the Milken study benchmarked Arkansas against other states in S&T, and ranked the state 49th, ahead of only Mississippi. Arkansas ranked 50th in categories such as competitive National Science Foundation funding and federal R&D. It was 49th in human capital and S&T workforce subsidies, and 49th for percent of the adult population with college degrees. Arkansas ranked 42nd in indices for risk capital and entrepreneurial infrastructure (leading other states in its region), and 12th in the nation for high tech industries' annual growth rate. Arkansas also ranked 25th in the Small Business Survival Index. The study observed that "Arkansas' strong performance in the area of net formation of high-tech establishments is very good news for the state's economy" but warned that if factors such as the quality of primary education did not improve, "Arkansas' competitiveness in the attraction and retention of high-tech firms will quickly erode."[37]

On the basis of the Milken study, members of Accelerate Arkansas concluded that the state needed to take a new approach to economic growth that was not dominated by efforts to recruit traditional manufacturing companies. In addition to the Milken study, Accelerate Arkansas studied the development plans of other states and deliberated extensively, both internally and with stakeholders throughout the state.[38] Based on this analysis, they identified five "core strategies for acceleration:"

---

[36] De Vol, et al., *Arkansas' Position in the Knowledge-Based Economy: Prospects and Policy Options* (2004) p. 1.
[37] De Vol et al., (2004) op. cit. pp. 3-4.
[38] The study of other states' programs was based on an examination of "best in class states," asking the question what makes some states "best" and how did they get that way? Between October 2005 and June 2006, Accelerate Arkansas' Strategic Planning Committee consulted 95 stakeholder groups in discussions over objectives for a strategic plan and held public strategic planning fora in various parts of the state. A series of 65 strategic objectives was winnowed down to 30. Through this process, five "core strategies" were identified. Teresa McLendon (ed.) *Building a Knowledge-Based Economy in Arkansas: Strategic Recommendations by Accelerate Arkansas* (September 2007) pp. 33-35.

1. support for job-creating research;
2. development of risk capital that is available for all stages of the business cycle, particularly the funding gap;
3. encouragement of entrepreneurship and new enterprise development;
4. increasing Arkansas' education level in science, technology, engineering and math;
5. and support for existing industry through improved technology and competitiveness.[39]

Arkansas Strategic Plan

In the wake of the Milken study, Governor Mike Beebe authorized the AEDC to develop a long range statewide economic development plan. In June 2007, AEDC Executive Director Maria Haley established a five person committee to create the plan. Input was solicited in the form of interviews and surveys from economic development stakeholders, all state agencies and commissions involved in economic development, the entire federal delegation, the governor's staff, and private businesses and foundations. The Committee drew upon the various studies and recommendations produced by Accelerate Arkansas, the Task Force for the 21st Century Economy, and the Governor's Workforce Cabinet. The Committee identified three basic challenges facing the state:

- The state needed a transitional, systematic approach to an economy based on knowledge-based jobs.
- Economic development efforts in Arkansas were diffuse and inefficient.
- Arkansas lacked a recurring and predictable funding formula for economic development.[40]

Governor Beebe's Strategic Plan concluded that the state's economic development efforts should be concentrated under the leadership of the AEDC, a decision based on overwhelming feedback from the interviews conducted by the AEDC committee. The plan tasked AEDC with developing a "recurring and predictable funding source." The Plan set forth five objectives:

- Increase the income of Arkansans at a rate faster than the national average,
- Expand entrepreneurship that focuses on knowledge-based enterprise,
- Prepare Arkansas businesses to compete more effectively in the global marketplace, Develop economic development policies that meet special

---

[39] McLendon (2007) op. cit. pp. 26-28.
[40] Governor Mike Beebe's *Strategic Plan* (2009).

needs and take advantage of existing assets in various areas of the state, and
- Increase the number of workers with post-secondary training.

Watt Gregory of Accelerate Arkansas pointed out at the conference that Governor Beebe's Strategic Plan is closely aligned with recommendations originally developed by the Task Force for the 21st Century Economy. These were to:
- Emphasize human resource development, particularly STEM education at all levels and workforce education
- Develop mechanisms to carry innovation into the marketplace, including:
- Support for entrepreneurship
- Additional risk capital
- Increased global competitiveness in recruiting businesses and industries
- Develop cyberinfrastructure
- Support innovation by existing businesses.[41]

Continuing Erosion of Jobs

Arkansas has continued to experience a net loss of manufacturing jobs. The manufacturing workforce declined from 269,815 in 2001 to 199,015 in September 2011, a drop of 25.9 percent for the decade.[42] In his conference keynote, Governor Beebe acknowledged that during the economic downturn that began in 2008 the state "lost jobs, more than we've gained — about 24,000 created, 27,000 lost," but pointed out that of the 27,000 job losses, many required less education and resulted from movement offshore or consolidation, whereas the new jobs were better-paying and required higher levels of education and skill. He emphasized that education and economic development must be linked or neither can succeed. "[Y]ou can't have economic development today without education, because you have to have the high-quality workforce. But

---

[41] See the summary of the presentation by Watt Gregory in the Proceedings chapter in this volume.
[42] The net decline has continued despite strong job growth in the oil and gas industry as a result of the development of major hydrocarbon deposits into the state. In 2004, Southwestern Energy successfully drilled test wells for natural gas in the Fayetteville Shale Play, a 20 by 100 mile deposit in north central Arkansas. A rush of development followed by Southwestern and other companies. A 2007 study by the University of Arkansas' College of Business calculated that in that year, exploitation of the Play would add $1.6 billion to the state's economy and employ 6, 000 people. Glen R. Sparks, "Community Profile: Conway, Ark. Makes Play for Economic Boom," *The Regional Economist* (July 2007).

without economic development, educated people may leave for other states or countries, in some instances for work."[43]

The Battelle Study

In 2007, the Arkansas General Assembly, implementing Accelerating Arkansas' core strategy of "support jobs-creating research, approved funding to support the establishment of the Arkansas Research Alliance (ARA),[44] a public-private partnership which seeks to foster university-based job-creating research. ARA sponsored a study by the Battelle Technology Partnership.[45] The study identified 18 "core competencies" in the state as well as 12 "niche competencies," winnowing those down to nine "strategic focus areas" – multidisciplinary fields of research likely to enable Arkansas to engage multiple institutions and leapfrog traditional universities with strengths in narrow academic fields.[46] Table 2 details the focus areas.

---

[43] See the summary of the keynote speech by Governor Mike Beebe in the Proceedings chapter in this volume.
[44] ARA funding was provided through the Arkansas Science and Technology Authority.
[45] Battelle Technology Partnership Practice, "Opportunities for Advancing Job-Creating Research in Arkansas, A Strategic Assessment of Arkansas University and Government Lab Research Base," 2009. Access the report at http://www.aralliance.org/__data/assets/pdf_file/0017/1682/Job-Creating-Research-in-Arkansas.pdf The 2009 Battelle study was based on interview with 85 of the top researchers in the state and quantitative revision based journal publications and research grants of faculty members during the preceding five years.
[46] See the summary of the presentation by Jerry Adams in the Proceedings chapter in this volume.

TABLE 2 Arkansas' Nine Strategic Focus Areas

Strategic Focus Area	Breadth of Competencies and Institutions	Opportunity for External Research Funding	Market Potential	Existing or Emerging Industry Connections
Enterprise Systems Computing	Emerging	Moderate	Extensive (immediate)	Extensive
Distributed Energy Network Systems	Emerging	Limited	Extensive (near term)	Extensive
Optics and Photonics	Emerging	Limited	Moderate (immediate)	Limited
Nano-Related Materials and Applications	Established	Significant	Extensive (long term)	Moderate
Sustainable Agriculture and Bioenergy Management	Established	Limited	Moderate (near term)	Moderate
Food Processing and Safety	Established	Moderate	Moderate (immediate)	Extensive
Personalized Health Research Sciences	Emerging	Moderate	Extensive (longer term)	Limited—addresses major public health issues
Behavioral Research for Chronic Disease Management	Emerging	Significant	Limited (immediate)	Limited – Addresses major public health issues
Obesity and Nutrition	Emerging	Significant	Extensive	Limited – addresses major public health issues

SOURCE: Battelle 2007.

The Battelle study offered the ARA a "crucial roadmap" to use in recruiting talent into the state and into the core focus areas, and that the ARA would use it as its "investment roadmap going forward."[47]

INITIATIVES TO PROMOTE INNOVATION IN ARKANSAS

The Arkansas Strategic Plan summarized the work of the various task forces and consulting firms that have analyzed and made recommendations on the state's economic growth. It noted that "unfortunately, many of the recommendations contained with these early reports were never fully implemented. As a result, numerous problems identified as early as 1964 still remain today."[48] Currently, however, there are a number of state, local and federal initiatives under way to address issues of concern raised by the Milken study and other reports released over the past decade.

Innovation Clusters

The 2004 Milken study of the Arkansas economy observed that "where clusters of existing technologies expand and emerging science-based technologies form is a critical factor in determining economic winners and losers in the first half of the 21st century." It observed that knowledge is generated, transmitted and shared more efficiently in close proximity, and economic activity based on new knowledge has a high propensity to cluster within a geographic area. The study pointed out that regional and state economic viability depends upon the ability to establish local technology clusters networked into the global business community. It concluded that clusters represent a state or region's best defense against "being arbitraged in the global cost-minimization game" and can "mitigate input-cost disadvantages through global sourcing."[49]

The Milken study analyzed various parameters supporting the formation of technology clusters and ranked Arkansas 44th among the states for technology concentration and dynamism. The study found that Arkansas lagged the national average for net annual formation of high technology establishments by 43 percent and trailed the 50-state average for proportion of the work force in high tech sectors by 64 percent. The state ranked 47th in start-ups of high tech companies. The state was strongest in high tech industries' average yearly

---

[47] See the summary of the presentation by Jerry Adams in the Proceedings chapter in this volume.
[48] Governor Mike Beebe's *Strategic Plan* for Economic Development, Little Rock: Arkansas Economic Development Commission, 2009. p. 6.
[49] Clusters are increasingly acknowledged as key to growing 21st Century innovation economies. See National Research Council, *Growing Innovation Clusters for American Prosperity.* C. Wessner, rapporteur, Washington, DC: National Academies Press, 2011.

growth, a category in which Arkansas exceeded the national average by 25 percent.[50]

A number of speakers at the conference highlighted the role that a number of state and federal initiatives can play to promote technology clusters in Arkansas.

Arkansas' Science and Technology Research Parks

In his conference presentation, Jay Chesshir of the Little Rock Chamber of Commerce noted that the state has two science parks, with a third on the way: The first is the Arkansas Research and Technology Park (ARTP), adjacent to the University of Arkansas in Fayetteville. As the state's primary knowledge community, the Fayetteville area provides valuable fuel for the innovation and technology development activities of the ARTP. ARTP's goal is to nurture technology-intensive companies through the formation of a community of companies, faculty and students drawing on "a set of core R&D competencies at the university." ARTP is managed by the University of Arkansas Technology Development Foundation. It is a 501(c)(3) organization with a mandate to "validate, develop and transfer inventions made at the University to Arkansas companies and start-up ventures."[51] The ARTP features the "Enterprise Center," a technology incubator for startups specializing in information technology and assembly manufacturing.[52] He added that two affiliates had received the prestigious Frost and Sullivan Award for excellence in technology, and another affiliate won the Tibbetts Award for the most innovative small business. Recently, another affiliate won an R&D 100 award. ARTP affiliates, he said, continue to advance the frontier of product development in many specialty areas.

A second park is the Arkansas Bioscience Innovation and Commercial Center at Arkansas State University in Jonesboro, which is completing its Phase I business incubator. The development of a third S&T Park in Central Arkansas accelerated in 2010 when the Arkansas legislature authorized the establishment of research park authorities.[53] A Tech Park located in Little Rock between UAMS (the University of Arkansas for Medical Sciences) and UALR (the University of Arkansas at Little Rock) is planned and part of the funding for the Technology Park will come from a voter approved increase in sales tax in Little

---

[50] De Vol et al., (2004) op. cit., p 50.
[51] ARTP, website, http://www.vark.edu/ua/artp/aboutus.html
[52] Financial support for the Center is being provided through an economic infrastructure fund grant and from the Arkansas Economic Development Commission. Arkansas Small Business and Technology Development Center, "Enterprise Center to Offer Valuable Technology Incubator Resources," press release, 2009, http://www.asbtdc.org/document/master.aspx?doc=1137
[53] Arkansas Act 1045 of 2007.

Rock. The revenue is estimated at $22 million of bonding capacity, with additional private matching funds being sought. Initial engineering evaluations of possible sites are now underway.

Arkansas Biosciences Institute

In her conference remarks, Carole Cramer of the Arkansas Biosciences Institute located at Arkansas State University observed that some 200 food processing facilities are located in the state, including those of some of the world's largest: Tyson Foods, Frito-Lay, Butterball, Wal-Mart, Riceland, Post, Nestle, and others. Wal-Mart alone, she said, "brings a cluster of people who want to sell to them." These firms, however, while they do their manufacturing and processing in the state, do most of their R&D elsewhere. Maintaining Arkansas' agricultural leadership, she said, will require will require building a significant in-state capacity for research and technological innovation.

To address this challenge, Arkansas State University at Jonesboro announced in 2011 the formation of a commercial innovation technology incubator at its Arkansas Biosciences Institute, to be known as the Arkansas Biosciences Institute Commercial Innovation Center (ABI-COM). ABI-COM will provide office and laboratory facilities for businesses seeking to turn innovations into products and services.[54] In his conference remarks, Barry Johnson, then of the Economic Development Administration of the U.S. Department of Commerce, noted that the agency has awarded a $1.75 million public works grant to Arkansas State University at Jonesboro and helped establish the Arkansas State Biosciences institute Commercial Innovation Center.

University of Arkansas for Medical Sciences (UAMS) BioVentures

Speaking at the conference, Michael Douglas of UAMS said that the objective of his organization was "building deals for Arkansas," and that he would offer a picture of UAMS BioVentures by touching on "the numbers, the process, best practices, state incubators, and results." UAMS BioVentures serves a dual purpose in driving the commercialization of life sciences in Central Arkansas. First, it works within the University of Arkansas for Medical Sciences to capture the intellectual assets of the research scientists in the form of patents, them finds buyers or licensees for those patents for commercialization. This yields license agreements with industry for early stage start-up companies. Second, BioVentures operates a mixed-use wet lab life sciences incubator that provides resources to UAMS start-up companies to assist their early-stage

---

[54] Arkansas State University, "Brian Rogers Named Director of Commercial Innovation Technology Incubator," press release, January 5, 2011. http://asunews.astate.edu/BrianRogersNamedABI-COM.htm ASU developed ABI-COM with support from a grant of $1,750,500 from the U.S. Department of Commerce Economic Development Administration. Ibid.

development into operating companies in the region. BioVentures manages the prosecution of a patent estate of about 250 patent cases, about 65 percent of which are issued or allowed. The incubator works closely with university start-up companies to assist their early-stage development by providing office and laboratory space. It also provides leads to technical support and management as well as a network of seed and venture funds to bring working capital to these early companies.

Dr. Douglas said that the economic productivity of the incubators was high, with an average annual wage of $56,000. The total capital raised by the incubators in Arkansas was $247 million, and the number of jobs created was 1252. He said that a 2009 economic impact study by the Institute for Economic Advancement found that BioVentures had initiated 44 company start-up projects since its inception; generated $29.4 million in revenues (in 2008) from new products, services, and research; and created $52.4 million (in 2008) in economic output impact, with 13 percent of the total out of state. Overall, the study found that the total economic impact (1997-2008) in sales, investment, and research in the state was $184 million.

USEDA Regional Innovation Clusters (RICs)

Speaking at the conference, Barry Johnson further noted that the U.S. Economic Development Administration is planning to staff its regional offices with personnel dedicated to RICs. In 2009, USEDA made 14 investments in Arkansas to support planning and implementation efforts to encourage clusters and regionalism in the state.[55] He defined RICs as "geographically bounded, active networks of similar, synergistic or complementary organizations that leverage their region's unique competitive strengths to create jobs and broader prosperity." Jobs within clusters pay higher average wages, and regional industries based on place-based advantages are less susceptible to off-shoring.[56]

In addition to the grant to Arkansas State University at Jonesboro noted above, Mr. Johnson noted that recent USEDA activities in Arkansas include over a dozen investments to support planning and implementation efforts aimed at encouraging regionalism and clusters across the state. This includes, he added, a recent USEDA Technical Assistance Grant to help establish the Center for Regional Innovation at the University of Arkansas at Little Rock.

---

[55] An USEDA Technical Assistance Grant helped establish the Center for Regional Innovation at the University of Arkansas at Little Rock. A $1.75 million USEDA public works grant to Arkansas State University at Jonesboro helped establish the Arkansas State Bioscience Institute Commercial Innovation Center. See the summary of the presentation by Barry Johnson in the Proceedings chapter in this volume.

[56] See the summary of the presentation by Barry Johnson in the Proceedings chapter in this volume.

Cyberinfrastructure

"Cyberinfrastructure" refers to the technological infrastructure that enables scientific inquiry, and includes high performance computing, data storage systems, data repositories and advanced instruments, visualization technology and people, all linked by advanced networks.[57] Dr. Amy Apon, until August 2011 the Director of the Arkansas High Performance Computing Center and Professor of Computer Science at the University of Arkansas, led an effort to establish and upgrade the state's cyberinfrastructure by securing federal and state support.[58] In 2007, Dr. Apon launched an effort to bring an outside team of experts, the External Advisory Committee, to study Arkansas' cyberinfrastructure needs, the result of which was a recommendation that the state launch the Arkansas Cyberinfrastructure Initiative.[59] In May 2008, Governor Beebe funded the Initiative through the Arkansas Science and Technology Authority, and an Arkansas Cyberinfrastructure Strategic Plan was drawn up in 2008 by members of research organizations in the state.[60]

In her conference remarks, Dr. Apon said that underlying the cyberinfrastructure initiative is the recognition that "computing has become the most important general-purpose instrument of science."[61] Research in fields such as nanotechnology, materials science, and human biology sometimes requires millions of hours of computing time per year.[62] Arkansas deployed a major new cyberinfrastructure resource in 2008, the "Star of Arkansas," the most powerful computer in the state, capable of storing over five times the data stored

---

[57] *Arkansas Cyberinfrastructure Strategic Plan* (2008).p. 3.
[58] In 2004, 2007 and 2010 Apon led efforts to secure Major Research Instrumentation (MRI) award from the National Science Foundation to establish a new campus research network and upgrade the power and chilling infrastructure of the university's data center. Clemson School of Computing, "Dr. Amy Apon Joins the School of Computing as Chair of the Computer Science Division," undated press release, http://www.clemson.edu/ces/computing/news-stories/aapon.html
[59] The External Advisory Committee was partially funded by "NSF Other" funds contributed by the University of Arkansas and the Arkansas Science and Technology Authority. The committee was chaired by Stan Anhalt, Director of the Ohio Supercomputer Center. The committee's work was supported by an Internal Steering Committee comprised of about 30 Arkansas leaders from the education, industry and public sectors.
[60] The Strategic Plan was drafted by the Arkansas Cyberinfrastructure Advisory Committee, which was similar in membership to the Internal Steering Committee.
[61] Jay Buisseau, Director of the Texas Advanced Computing Center, cited by Amy Apon, in her remarks at the conference. See the summary of her remarks in the Proceedings section of this volume.
[62] Dr. Amy Apon cited research underway by Professor Peter Pulay at the University of Arkansas to study the interaction of chemicals on human protein and DNA that requires four million hours of computing per year. Assistant Professor Doug Spearot is creating three-dimensional models of alloys that do not yet exist, using 20 million or more atoms, an effort which requires six million hours of computing time per year. Nanotechnology device research by Professor Laurent Ballaiche requires 70 million hours of computing time per year. See the summary of the presentation by Dr. Apon in the Proceedings chapter in this volume.

in the entire Library of Congress.[63] This infrastructure now includes the research universities of Arkansas and the four-year colleges.

CHALLENGES

Despite substantial progress in the development of a knowledge-based economy, a number of speakers pointed out that Arkansas confronts key challenges in developing a cadre of educated, skilled labor, in providing adequate venture capital for start-ups, and in securing more federal research funding. Several speakers at the conference drew attention to these challenges.

Building Human Capital

Arkansas' labor force has won frequent praise for its work ethic, and a number of major manufacturers who have chosen to locate in the state have cited the motivation of the work force as an important factor for locating there.[64] However, with an increasing need for highly educated and skilled workers, Arkansas has found that it trails other states and regions in most indices for assessing skilled human capital. In 2006, the state ranked last among the 50 states in percent of adults with college degrees. A number of executives from technology-intensive companies noted at the conference that this shortage of talent in the state requires them to establish costly training programs when they

---

[63] Purchase of the Star of Arkansas was funded in part by an $803,306 grant from NSF, with matching funds from the University of Arkansas and in partnership with Dell Corp. Stan Anhalt, Chair of the External Advisory Committee, observed that the Star of Arkansas had "the potential to improve Arkansas' economic future through research in areas such as natural gas production, bird flu prevention, rice irrigation, nanotechnology, large-scale transportation and commerce systems, material design, sustainability, and personalized medicine. The Star of Arkansas is eight times faster than the University of Arkansas' other supercomputer, Red Diamond, which was acquired in 2005. University of Arkansas College of Engineering. "University of Arkansas Installing Supercomputer, 'Star of Arkansas', to be State's Fastest," press release, 2008, http://www.engr.engr.vark.edu/home/2378.php

[64] Nucor Steel has located a number of facilities in Arkansas where local hires were typically "farmers or machinery workers who have been ingrained with a strong work ethic since childhood." Dan DiMicco, President of Nucor-Yamato Steel Co., commented on his operation in the state that "We hire good people, put them in a culture that encourages them to do well, give them the tools and the opportunity to excel and then we get the heck out of their way." "Nucor Makes Blytheville Steel Capital of the South," *Arkansas Business* (December 16, 1996). Mitsubishi Power Systems Americas, Inc. cited Arkansas' "extraordinary work ethic" as a factor underlying its decision to locate a $100 million manufacturing facility for wind turbines in Fort Smith, Arkansas. The plant's general manager said that "we looked for a part of the country where manufacturing is not some lost art." "Mitsubishi Breaks Ground on Nacelle Facility in Arkansas," *North American Windpower* (October 8, 2010).

established businesses there.[65] Their remarks echo a 2006-07 survey of Arkansas business leaders where 76 percent of those surveyed said that more than half of job applicants who recently graduated from high school lacked basic math and writing skills.[66]

The Arkansas Task Force on Higher Education Remediation, Retention and Graduation Rates was formed pursuant to legislation enacted by the General Assembly in 2007 to study the state's education system with an eye toward increasing the percentage of citizens with bachelor's degrees. At that time, the percent of adults in Arkansas holding bachelors' degrees was 22.3 percent – well below the average for the 16-state Southern Regional Education Board (SREB). While Arkansas exceeded many Southern Regional Education Board (SREB) states in the number of high school graduates entering college, a greater percentage of those entering college failed to complete bachelors' degree programs.[67]

Reforming of the Public Schools

Governor Beebe and other conference speakers listed the steps that the state is taking to improve PreK-12 education. Previously, only two counties in the state produced college entering populations with the percentage of students requiring remedial math lower than 25 percent. At the University of Arkansas at Pine Bluff, the percentage of entering freshmen in 2007 requiring remedial courses was 75.5 percent for English, 84.9 percent for math, and 73.6 percent for reading.[68] A 2006-07 survey of Arkansas college professors on the overall

---

[65] See, for example, the summary of the presentation by Jeff Johnson in the Proceedings chapter in this volume. Nordex USA found that when it sought to open a manufacturing plant for wind power equipment in Arkansas, it was only able to find after five months of interviews 62 of the estimated work force of 700 who possessed sufficient skills. Because of the high level of skill required, Nordex plans to build a training academy onsite. See also the summary of the presentation by Joe Brenner in the Proceedings chapter in this volume.

[66] Arkansas Department of Education, *Combined Research Report of Business Leaders and College Professors on Preparedness of High School Graduates* (January 2007). Similarly, the human resources manager at Kagome/Creative Foods, a food processor with a facility in Mississippi County, Arkansas, said in 2010 that despite the county's high unemployment, it was "very, very hard to find people to work," partly a case of "too many undereducated, unemployable youth." Susan C. Thompson, "Factory Closing Shocks Community into Opening Wallets for Economic Development," *The Regional Economist* (October 2010).

[67] At the time the Task Force was formed, of every 100 students then in ninth grade, 74 would graduate from high school, 64.7 would enter college, and only 16 would graduate with an associate or bachelors' degree within 10 years. Of 37,160 students who graduated from ninth grade in 1996, only 5,817 had achieved higher degrees by 2006. The Task Force warned that "The pipeline is broken. Can a modern economy be built upon 5,817 people?" Arkansas Task Force on Higher Education Remediation, Retention and Graduation Rates, *Access to Success: Increasing Arkansas' College Graduates Promotes Economic Development* (August 2008) ("Education Task Force Report.")

[68] Education Task Force Report (2008) pp. 11, 14.

academic quality of the Arkansas public high schools in preparing students for college found that over half gave the public schools grades of D (50.2 percent) or F (9.6 percent).[69]

A decade ago, the shortcomings in the state's PreK-12 educational system were sufficiently severe that the Arkansas courts declared the state's system of school funding to be "inadequate under ... The Arkansas Constitution." The Chancery Court stated that "Too many of our children are leaving school for a life of deprivation, burdening our culture with the corrosive effects of citizens who lack the education to contribute." The court declared that under the state's constitution, financing must be based on the amount of funding required to provide an "adequate educational system," and ordered a cost study.[70] The Arkansas Supreme Court upheld the lower court's decision in 2002. The net result of the Lake View decision was a substantial increase in state funding for operations and facilities in elementary and secondary schools as well as overhaul of the curriculum, increased teachers' salaries and increased requirements for accountability from school districts. By the time of the Education Task Force's report, these reforms were beginning to have positive effects.[71] As Governor Beebe noted at in his keynote speech at the conference, Arkansas is now winning accolades for levels of per-pupil funding, test scores, transparency, accountability, standards, and increase in advanced placement students.[72]

The Arkansas Science Technology, Engineering and Math (STEM) Coalition is a statewide partnership of leaders from the business, government, education and community sectors to develop and implement policies to improve STEM teaching and learning. In his conference presentation, Michael Gealt of the Arkansas STEM Coalition noted that his organization is concerned with all levels of education from pre-K onward, and functions both as a think tank for ideas for improving STEM education and as a lobbying organization seeking public policies to improve STEM education. Among other initiatives, the coalition has secured funding for 27 elementary school science specialists, sought state grants to STEM teachers to increase their income, established a web

---

[69] Arkansas Department of Education, *Combined Research Report of Business Leaders and College Professors on Preparedness of High School Graduates* (January 2006).

[70] Lake View School District No. 25 v. Huckabee No. 1992-5318 (Pulaski County Chancery Court), May 25, 2001.

[71] In 2008, over half of Arkansas students scored "proficient or above" on the state's tests for mastery of grade-level knowledge, whereas proficiency rates a decade previously were between 20 and 40 percent. In 2007, U.S. Secretary of Education Margaret Spellings cited Arkansas and Massachusetts as two states with education reform models that other states should emulate. Education Task Force Report (2008) p. 12

[72] See the summary of the keynote speech by Governor Mike Beebe in the Proceedings chapter in this volume.

portal for STEM lesson plans, and advocated differential pay for STEM teachers.[73]

Training and Retaining University Graduates

Arkansas has 11 four-year institutions of higher learning, two of which have research as a fundamental part of their missions, the University of Arkansas, Fayetteville and the University of Arkansas for Medical Sciences.[74] Major public and private commitments have been undertaken to improve the quality of these institutions, efforts that are being reflected in a succession of national research awards and individual faculty and student achievements. In addition, the University of Arkansas at Little Rock is now classified as a Research University with high research activity, and Arkansas State University at Jonesboro is classified as a Doctoral Research University. These four constitute the "Research Universities" of Arkansas.[75]

The fact that Arkansas currently ranks 50th in the U.S. in the percentage of adults with a college degree is not necessarily an indication of the failure of the state's universities; the problem is one of an out-migration of graduates. The 2009 Battelle study found that Arkansas' university system increased the number of graduates from 12,153 in 2001 to 15,262 in 2007, a growth of nearly 26 percent. Graduates in fields related to science, math and engineering grew from 3,548 in 2001 to 4,341 in 2007, an increase of 22 percent. Most significantly, the number of doctorate degrees in science, math and engineering from Arkansas doubled during the same period, from 65 doctoral degrees to 130, a growth rate nearly double the national average, suggesting that the state's "strong growth in research funding is translating into top-level talent creation in the state." Health and clinical sciences led the growth in doctoral degrees.[76]

University of Arkansas Chancellor David Gearhart recently pointed out that while the state needs to graduate college students at a higher rate, in fact between 1989 and 2006 Arkansas produced 166,000 college graduates, "nowhere near the last place nationally." Arkansas has ended up at the bottom of the rankings because during the same period, 42 000 of those graduates left the state, with the most beneficiaries of this migration being "states with human capital economies." As the Chancellor framed the challenge facing the state, "if

---

[73] See the summary of the presentation by Michael A. Gealt in the Proceedings chapter in this volume.
[74] More information on assets, reach, and research expertise of the University of Arkansas system can be accessed at http://www.uasys.edu/. More information on the Arkansas State University system can be accessed at http://www.asusystem.edu/.
[75] For some considerations, the University of Arkansas at Pine Bluff is considered as a fifth research university because of its research on aquaculture.
[76] Battelle study (2009) op. cit., p. 5

graduates are leaving to go where those businesses already are, how do you reverse the process and attract these businesses to your region?"[77]

Speaking at the conference Jeff Johnson, the CEO and president of ClearPointe (a managed service provider headquartered in Little Rock), described the challenges he faced in finding local IT talent to staff his Arkansas-based company. At first the company tried "growing their own" – hiring new college graduates and putting them through six months of training before placing them in company roles. This proved to work well, he said, but at a high cost for a small company, with a significant number of employees tied up in training for extended periods. Hiring talent from out of state also was difficult.

In this regard, he noted, the creation of the Engineering and Information Technology College at UALR was a "godsend." "We started working with the college more than three years ago, serving on the advisory council." In the past year, Mr. Johnson and others have helped the college design a curriculum that would assure companies like ClearPointe of a steady flow of qualified applicants.

Research funding in Arkansas universities has been growing rapidly, but from a comparatively weak starting point – in effect the state is still playing catch-up. The Battelle study found that in 2007 spending on university-based R&D in Arkansas totaled $240 million, an amount equal to 0.25 percent of gross state product (GSP), and that to achieve the national average of 0.36 percent of GSP, spending would have needed to be $106 million higher. The level of funding for university research in the state had grown 70 percent between 2001 and 2007, exceeding the national average growth rate of 51 percent.[78] However, the rapid growth in Arkansas university research was evident across the research spectrum (see Table 3).

TABLE 3 Growth Rate in Arkansas University R&D Funding, 2001-2007

Field	Arkansas Growth Rate (percent)	National Average
Biological sciences	133	55
Physics	94	31
Chemistry	205	44
Other engineering	105	28
Other life sciences	443	50.5

SOURCE: Battelle Study, 2009, p. 3.

---

[77]Gearhart, David. "Arkansas 180: Teaching & Research," http://chancellor.vark.edu/13132.php.
[78] Battelle Study (2009) op. cit., pp. vi-viii.

Addressing the Venture Capital Gap

A region's ability to foster innovative start-up companies is critical to its success in the knowledge-based economy. At present Arkansas' ability to generate start-ups is being constrained by the shortage of venture capital financing to enable new companies to reach the commercialization stage for new technologies. As Richard Bendis noted in his presentation, many small start-ups perish for lack of funding before they can commercialize their products, a primary factor underlying the so-called "Valley of Death" phenomenon. During the early stages of product development, start-ups need access to capital, such as angel or venture financing. However, most angel investments are very small and venture capital investments have moved downstream, toward established firms already generating revenues and profits. The current average venture capital investment is $8.3 million, signifying the move away from risky start-ups that typically require only a fraction of that amount.[79]

The Task Force for the Creation of Knowledge-Based Jobs said in its 2002 Report that "A key element that has been missing from the entrepreneurial equation in Arkansas is the lack of venture capital to keep new, knowledge-based businesses in the state."[80] Nearly a decade later, as Dr. Mary Good observed in her conference remarks, the state has two pressing needs — to improve the access to very early state capital for start-up firms, and to raise sufficient funding for innovative initiatives.[81]

The Milken study found that an average of about one (0.96) Arkansas firm per 10,000 businesses received venture capital from 1993 to 2002, a rate which was one-sixth the national average. Between 2002 and 2004, the number of state firms per 10,000 businesses receiving venture capital tripled, an increase that indicated that "venture capitalists were beginning to discover the state."[82] Nevertheless, Arkansas has yet to develop strong links with private equity. Venture capital investments in the state prior to 2006 usually totaled under $10 million per year, and after a spike to $40 million in that year fell off to nearly zero in 2007-2008. Even the 2006 total of $40 million, the state's best year for venture investment represented only 0.15 percent of U.S. venture capital investments in that year.[83]

In his conference remarks, Jeff Johnson, said that ClearPointe's experience as an IT start-up in Arkansas underscores the challenges confronting local high tech start-ups. The company's only original source of capital was its receivables. Seeking financing, in 2002 the company was chosen as a presenter

---

[79] See the summary of the presentation by Richard Bendis in the Proceedings chapter in this volume.
[80] *Report of the Task Force for the Creation of Knowledge-Based Jobs* (September 2002). P. 26.
[81] See the summary of remarks by Dr. Mary Good in the Proceedings chapter in this volume.
[82] De Vol et al., (2004), op. cit. p. 81.
[83] See the summary of the presentation by Richard Bendis in the Proceedings chapter in this volume.

at the Arkansas Venture Capital Forum, but virtually no funding was available for IT start-ups in the wake of the bursting of the dot-com bubble. The Forum gave ClearPointe access to knowledgeable people who helped it refine its business plan and monetize its needs. This attracted financing from local banks and, as Johnson recounts, "bank loans are not the best way to start a company, but we had no other options." The company was able to operate on a "pay-as-you go" basis until it could attract some angel backing, which in turn facilitated bank loans. Jeff Johnson identified access to funding as the company's highest hurdle to overcome.[84] And, as the Arkansas Strategic Plan points out, recent structural changes in the banking industry make it more difficult to obtain the kind of debt financing that sustained ClearPointe through its early stage development.[85]

According to Watt Gregory, chair of the Executive Committee of Accelerate Arkansas, a number of public and private institutions in Arkansas are working to extend financial support to start-up companies in the state. The state provides funding for extremely early stage companies through AEDC's Targeted Business Tax Incentives and the ASTA Seed Capital Fund.[86] Arkansas Capital Corporation Group (ACCG) is a private, not-for profit entity comprised of several affiliated companies dedicated to financing economic development in the state. The flagship company, Arkansas Capital Corporation (ACC), provides long-term, fixed rate loans to Arkansas businesses. Diamond State Ventures, affiliated with ACCG, provides venture capital investments ranging from $250,000 to syndicated investments up to $20 million. Another ACCG member, Commerce Capital Development Company, supervises investment tax credits provided pursuant to the Arkansas Capital Development Company Act.[87] Arkansas Certified Development Corporation, also an ACCG member, administers SBA 504 loans.[88] A group of angel investors created an $8 million Fund for Arkansas' Future that provides start-up funding in the $100,000 to $500,000 range.

The Arkansas Institutional Fund (AIF) is a fund-of-funds that invests in private equity and venture capital funds directing early stage capital, traditional venture capital, later stage and expansion capital and special situations capital

---

[84] See the summary of the presentation by Jeff Johnson in the Proceedings chapter in this volume.
[85] Governor Mike Beebe's *Strategic Plan* (2009) op. cit., p. 45
[86] The tax incentive program provides transferable tax credits to start-ups that do not have a state tax liability. The credits can be sold to individuals or institutions with tax liabilities. Governor Mike Beebe's *Strategic Plan* (2009).op. cit.
[87] These equity investment credits now reside with the Arkansas Economic Development Commission.
[88] See the summary of the presentation by Watt Gregory in the Proceedings chapter in this volume; See also the ACCG website, http://www.arkansaedc.com/bring-your-business-to-arkansas/financing/arkansas-capital-corporation.aspx accessed January 11, 2012.

investments toward Arkansas enterprises. It was established pursuant to legislation authorizing the Arkansas Development Finance Authority (ADFA) to implement a venture capital investment program.[89]

Synergies in Federal Innovation Funding

The 2009 Battelle study used the level of federal funding as an important metric in identifying Arkansas' areas of strength in research. According to the Arkansas Research Alliance's founder, Jerry Adams, federal funding is the accelerator. "State funding can help support talent, but federal funding is the key."[90] The 2004 Milken study found that "Arkansas receives approximately $44 per capita in federal money for research and development activities. For the year measured (FY 2000), Arkansas received $117.8 million in federal R&D, the least of any state and less than 1/5000th of the national total. Averaged out per person, this amount of funding ranks the state 50th in the nation."[91] The states' underperformance in securing available federal research dollars represents a significant handicap in the competition with other states for knowledge-based companies and jobs. Addressing this challenge, several speakers at the conference highlighted important sources of federal funding for innovation and encouraged the state to take full advantage of these federal programs.

The SBIR Program

The Small Business Innovation Research (SBIR) is a competitive innovation award program that encourages small businesses to develop innovative technologies that address a variety of federal missions. It is funded through eleven participating federal agencies whose R&D budgets exceed $100 million, with each agency contributing 2.5 percent of their extramural R&D budgets to SBIR programs. SBIR frequently provides the first funding to help small innovative companies start projects, including support for academic researchers who have no company affiliation. Often, the "certification effect" of an SBIR award can help attract private investment and increase the prospects for winning a public contract.[92] Charles Wessner noted in his conference

---

[89] Governor Mike Huckabee authorized the establishment of the Arkansas Venture Capital Investment Trust to hold the equity interest in the AIF. The trustees of this trust are the President of ADFA, the President of Arkansas Science and Technology Authority, and the Director of the Arkansas Department of Finance and Administration. AIF website, http://www.arkansasinstitutionalfund.com/aif/web.nsf/pages/history.html
[90] See the summary of the presentation by Jerry Adams in the Proceedings chapter in this volume.
[91] De Vol et al., (2004) op. cit., p. 61.
[92] Following the 2012 reauthorization of the SBIR program, Phase I SBIR awards were raised to $150,000. These are provided for feasibility and proof of concept research. Phase II awards were raised to $1 million. These are intended to develop prototypes or products that are ready for

presentation that a recent assessment by the National Academies concluded that SBIR was "sound in concept and effective in practice, and had a positive impact on small firm formation and growth."[93]

In her conference remarks, Carole Cramer of the Arkansas Biosciences Institute noted that Arkansas has many potential opportunities if it learns to combine its traditional strengths in agriculture and food processing with new techniques of biotechnology. She said that she had herself co-founded in 1993 a small innovative business, CropTech Corp., with SBIR funding and Advanced Technology Program (ATP) grants. The company grew to 42 employees and, in 1999, won a patent for the production of human lysosomal protein expressed in plants.[94] The shock of 9/11 brought the work to a halt, but in 2003 the technology was licensed to Protalix Biotherapeutics. "The valley of death is real," she said, "but a good idea can survive." She added that in December 2009, Protalix completed a deal with Pfizer, demonstrating the commercial potential of using plant cells to make protein-based drugs.

In 2011, Arkansas companies won SBIR awards for research in areas including nanotechnology, pharmaceuticals, medicine and microelectronics.[95] However, the total SBIR awards won by the state in 2011 (17) was lower than the total for 2004 (24). In that year the Milken study ranked Arkansas 49th among the 50 states for SBIR awards per 100,000, and 50th for Phase II SBIR awards per 10,000 businesses.[96]

The National Institute for Standards and Technology (NIST)

Mr. Marc Stanley, deputy director of NIST, expressed his enthusiasm for "trying to grow new companies that have difficulty in trying to find early-stage investment money." On a policy level, he said, the most significant need was for federal agencies to move beyond their restricted silos of activity to more collaboration with other agencies with similar objectives in accelerating innovation.

The key areas for collaboration, he said, were regional policy, economic and industry policy, education policy, and science and technology policy. For example, he said that NIST, through the Manufacturing Extension

---

commercialization or application. A Phase III for product development and commercialization has been discussed but not funded. SBIR recipients retain the intellectual property for technology developed through the program.

[93] National Research Council, An Assessment of the SBIR Program, C. Wessner, ed., Washington, DC: National Academies Press, 2008.

[94] The U.S. Patent and Trademark Office in 1999 issued patent 5,929,304 for "Production of Lysosomal Enzymes in Plant-based Expression Systems" to Carole Cramer.

[95] "Arkansas SBIR/STTR Grant Awards," Arkansas Small Business and Technology Development Center website, http://asbtdc.org/DocumentMaster.aspx?doc=2416

[96] De Vol et al., (2004) op. cit., p. 28.

Partnership (MEP), would have a vital but limited role in the energy regional innovation cluster. "We have to see if we can get out of our silos and help states like Arkansas keep their students here, grow companies, raise good revenue, and help our country grow." He said he was impressed with Arkansas' efforts to support its own innovation activities, especially in the face of a severe economic downturn.

In his conference presentation, MEP Director Roger Kilmer described his organization as a network of 440 service locations and 60 MEP centers that work with small and mid-sized manufacturers to promote technology acceleration, supplier development, workforce improvement, sustainability and continuous improvement of manufacturing process. In 1998, Arkansas Manufacturing Solutions (AMS) was established by the Arkansas Science and Technology Authority as an affiliate of NIST's MEP to provide manufacturing extension services to local manufacturers. MEP concentrates on helping existing small firms scale up based on a national perspective.[97] In Arkansas, AMS services reportedly have facilitated $592 million in new and retained sales, $25 million in capital investment, $12.7 million in cost savings, and 3,335 jobs retained and created. MEP has created an Arkansas-specific portal into the National Innovation Marketplace — "Arkansas Innovation Marketplace (AIM)" — that lists technologies, intellectual property, and the capabilities of entrepreneurs, inventors and companies in the state.[98]

The National Science Foundation

The National Science Foundation (NSF) accounts for about one fourth of all federal funds awarded to U.S. colleges and universities for basic R&D. The Milken study found that Arkansas tied for last place (47th) with respect to the NSF funding rate for proposals received.[99] However, as NSF's Donald Senich pointed out at in his conference remarks, NSF has established an Industry/University Cooperative Research Center (I/UCRC) for engineering and logistics in Arkansas, featuring collaboration between Sam's Club and the University of Arkansas. NSF made 15 grants to the University of Arkansas between 2002 and 2009.[100]

---

[97] MEP Center projects include business growth services, technology services to develop products and processes, "lean" manufacturing techniques to promote continuous improvement, quality systems and other standards, advice on energy and sustainability, and development of talent. See the summary of the presentation by Roger Kilmer in the Proceedings chapter in this volume.
[98] As of late, 2011 AIM had posted about 50 technologies and 50 company "needs and wishes." See the summary of the presentation by Roger Kilmer in the Proceedings chapter in this volume.
[99] De Vol et al., (2004) op. cit., p. 78
[100] See the summary of the presentation by Donald Senich in the Proceedings chapter in this volume.

PROMISING SECTORS

The National Academies conference highlighted the state's economic and technological challenges as well as promising growth areas. As we see below, several speakers noted that the development of these sectors would build on the state's current and potential competencies and leverage Arkansas' areas of strength to grow knowledge-based industries and jobs.

Electric Power

The Battelle study identified "energy network systems" as a potential strategic focus area for Arkansas. As Paul Suskie of the Arkansas Public Service Commission pointed out at the conference, the regulation of electricity is shifting from promoting electricity consumption to incentivizing energy efficiency and renewable energy.[101] Nick Brown, CEO of the Southwest Power Pool (SPP), an electric power cooperative, also noted at the conference that much of the transmission grid in the south central U.S. is nearing the end of its useful life, and will need to be replaced. These imperatives are opening up economic opportunities for Arkansas-based businesses in fields such as renewable energy and electric power transmission.[102] Arkansas already supports a number of research organizations and businesses, which give it advantages in developing the electric power sector.

University of Arkansas Research Center
The Battelle study noted that a core research competency in power electronics was emerging at the University of Arkansas campuses at Fayetteville and Little Rock. In his conference presentation, Alan Mantooth of the National Center for Reliable Electric Power Transmission (NCREPT) said that his organization was founded at the University of Arkansas, Fayetteville to accelerate development of technologies for the electrical grid. NCREPT employees and graduate students conduct industrially-relevant research into future energy systems, including power electronics, with emphasis on grid reliability, power interface applications, transportation, energy exploration and geothermal applications.[103] NCREPT operates as a testing, prototyping and

---

[101] See the summary of the presentation by Paul Suskie in the Proceedings chapter in this volume.
[102] See the summary of the presentation by Nick Brown in the Proceedings chapter in this volume
[103] NCEPT Executive Director Alan Mantooth defines power electronics as "the interface between where we've generated the power and how we want to condition that power specifically for the load." Examples include power converters that operate between an electric plug and the motor of an appliance or the hard drive of a computer, and "intelligent" lighting systems that turn themselves off when people are not present in a room. At present 30 percent of the electricity generated in the United States is processed by power electronics, a figure that is forecast to rise to 80 percent by

industrial collaboration center for global users, which include companies and universities.[104] It is the only facility in the world offering programmability and reconfiguration operations at six megawatts.

Southwest Power Pool

The Battelle study also identified the Southwest Power Pool (SPP) as an important asset in developing the states' potential in electric power transmission.[105] At the conference, Nick Brown described the Southwest Power Pool as a cooperative organization that was originally formed in Arkansas during World War II to ensure sufficient electricity to support production of aluminum for the war effort.[106] SPP was created because the aluminum plants' electric power needs exceeded the entire generating capacity of the state, necessitating the formation of a pool originally comprised of 11 regional utilities. Originally comprised of 11 utilities in the region, SPP has expanded to 56 members operating in nine states.[107] SPP manages the flow of power over electrical networks, operates as a wholesale sales agency for power and serves as a "one-stop shop" for the sale of transmission services.

Wind Energy

According to Joe Brenner of Nordex, a manufacturer of wind turbines, Arkansas is already "a manufacturing powerhouse for the wind industry," and has become a manufacturing base for some of the most competitive makers of wind equipment in the world.[108] Arkansas is located at the edge of the "Saudi Arabia of wind" – the U.S. great plains states – and its strategic geography has been cited by wind power equipment manufacturers as a key factor in their

---

2030. See the summary of the presentation by Alan Mantooth in the Proceedings chapter in this volume.

[104] NCREPT won an "R&D 100" award from R&D magazine for innovation in 2009 for the development of a 3"x5" power electronic module for hybrid electric vehicle motors. Current modules must be actively cooled by the radiator, but the NCREPT device can operate at 250 degrees C., does not require water-cooling, and is lighter and more resilient than existing models. The new module was developed with funding from Japan's Rohm Semiconductor and Sandia National Laboratory, and was manufactured in Fayetteville by NCREPT and Arkansas Power Electronics. See the summary of the presentation by Alan Mantooth in the Proceedings chapter in this volume

[105] *Battelle Study* (2009) op. cit., p. 21.

[106] At the beginning of the 1940s Arkansas had the largest commercially exploitable deposits of bauxite in the United States. Alcoa and the Reynolds Metal company established plants in the state, and a United States government entity, Defense Plant Corporation, built an aluminum factory in Jones Mill, Arkansas, which was leased to Alcoa.

[107] SPP members include utilities, cooperatives, state and municipal agencies, and bulk power marketers.

[108] Nordex USA, a subsidiary of Nordex SE, a German manufacturer of wind turbines and a pioneer in the development of wind-driven power generation.

decision to establish a presence in the state.[109] In 2010, the wind industry supported 1-2,000 direct and indirect jobs in the state.

In his conference remarks Joe Brenner cited the strong support for his company by state and local leaders and a positive environment for innovation. In 2008, Nordex USA selected Jonesboro, Arkansas as the site for a manufacturing facility for 2.5 megawatt wind turbines. The plant, which became operational in 2010-11, is one of the most technologically sophisticated facilities in its kind in North America. Nordex chose the Arkansas site because of the commitment of state and local leaders to economic development, the availability of a trainable work force, the nearby presence of Arkansas State University as a site for training programs, and Arkansas' central location in North America.[110]

U.S. Department of Energy Initiatives

The federal government is committing substantial resources to the promotion of energy efficiency and renewable energy (EERE). Dr. Gilbert Sperling of the U.S. Department of Energy (DOE) Office of Energy Efficiency and Renewable Energy noted that the American Recovery and Reinvestment Act (ARRA) had temporarily augmented his office's budget, normally around $2 billion, to $16.8 billion. Of this amount, $11.5 billion was returned to the states to stimulate building weatherization and other efficiency-enhancing measures.[111] DOE's EERE initiatives, he said, seek to increase the market share of renewable energy power generation from the current one percent to 30, 40, or 50 percent. This effort requires convincing the American people why EERE investments

---

[109] See the summary of the presentation by Joe Brenner in the Proceedings chapter in this volume. The sheer size and weight of large wind turbines make transportation costs a factor in locational decisions. The turbines manufactured in Jonesboro will be as tall as a football field is long, and each turbine blade will be a comparable length. See the summary of the presentation by Nick Brown in the Proceedings chapter in this volume; "Arkansas Wins $100 Million Wind Turbine Nacelle Plant," *Energy Overviews* (May 11, 2011); Interview with Joe Brenner, Mitsubishi Power Systems America, in *Wind Systems* (January 2011).

[110] The University worked with Nordex and Beckmann Volmer, a supplier to Nordex of turbine mainframes and other components, to create classes and degrees to meet the unique need of the wind power industry. The University now offers training in "mechatronics," a combination of electrical and mechanical skills specific to the manufacture of wind turbines. "Beckmann Volmer Breaks Ground on Osceola Plant," *Paragould Daily Press* (September 14, 2011); Interview with Joe Brenner, Vice President of Nordex USA, in *Wind Systems* (January 2011).

[111] Arkansas received over $117 million in grant money for energy efficiency and renewables projects. Another $34 million went to the state in the form of tax incentives for wind power generation, electric vehicles, batteries, and other renewable and energy efficiency technologies. See the summary of the presentation by Gilbert Sperling in the Proceedings chapter in this volume. Nordex USA received $22 million from ARRA in the form of tax credits to support its manufacturing facility in Jonesboro, Arkansas. "Firm Building Jonesboro Plant to Get $22 Million Stimulus," *NWA Online* (January 11, 2010).

are warranted, securing private capital investment in renewables, and the removal of incentives for utilities to make more money by selling more energy. Specific initiatives by EERE include an effort to have one million plug-in hybrid electric vehicles on the road by 2015; an improvement of the federal government's energy efficiency; "recover through retrofit" (RTR) initiative to promote home energy efficiency; the concentration of developmental block grants for residential and commercial energy efficiency retrofits; and the development of technologies for concentrated solar power, geothermal energy wind power, biofuels, and hydropower.[112]

AEDC Wind Study

Arkansas' own wind power generation capability is still underdeveloped.[113] As Joe Brenner noted in his conference presentation, "There were challenges to finding the right locations in some parts of the state," but "siting specialists are quite sure that Arkansas can provide wind energy."[114] In 2011, AEDC's Arkansas Energy Office commissioned a "tall tower" study of wind velocity at various points in the state to generate data to afford wind power developers at better sense of the availability of wind resources in the state. The Energy Office is providing a grant to fund a wind resource monitoring network comprised of sensors on existing communications towers at the 80-foot level, the hub height of standard industry wind turbines.[115]

Nanotechnology

Speaking at the conference, Dr. Salamo, a Distinguished Professor of Physics at the University of Arkansas, Fayetteville, defined nanoscience as "the effort to understand and design structures at the nano scale and to seek their application."[116] The Arkansas effort in nanoscience, he said, is a collaborative undertaking among partner institutions throughout the state university system.

Materials Research Science and Engineering Center

The National Science Foundation has funded the establishment of a network of Materials Research Science and Engineering Centers (MRSEC) at academic institutions across the United States to undertake materials research, develop human resources, and collaborate with industry in materials science. A

---

[112] See the summary of the presentation by Sperling in the Proceedings chapter in this volume.
[113] At the end of 2010, the state had 10 megawatts (MW) of wind power generating capacity online with another 210 MW in planned projects. According to an estimate by the National Renewable Energy Laboratory, Arkansas has sufficient wind resources to provide 58.3 percent of the state's current electricity needs. "Arkansas is a National Leader in Wind Energy Manufacturing," *American Wind Energy Association* (August 2011).
[114] See the summary of the presentation by Joe Brenner in the Proceedings chapter in this volume.
[115] "A Wind Study the Size of Arkansas," *Wind Power News* (April 1, 2011).
[116] See the summary of Professor Salamo's presentation in the Proceedings chapter of this volume.

MRSEC has been established jointly at the University of Arkansas and the University of Oklahoma to support an interdisciplinary research program on semiconductor nanostructure science and applications. The Arkansas/Oklahoma MRSEC is pursuing nanotechnology research with applications in energy efficiency, conversion of waste to electricity, solar power generation, semiconductor technology, medical diagnostics, and cancer treatment. The MRSEC has resulted in six spin-off companies. Dr. Salamo noted that the University of Arkansas System's schools lead the U.S. in the supply of nanomaterials to research organizations across the United States.

National Center for Toxicological Research

Arkansas' potential for developing nanotechnology appears to be particularly promising in the area of life sciences. As Watt Gregory noted at the conference, the U.S. Food and Drug Administration (FDA) operates the National Center for Toxicological Research (NCTR) near Pine Bluff, Arkansas. This facility employs 500 people, nearly half of whom are Ph.D. level researchers and scientists. NCTR research themes include food contaminants, detection of terrorist threats, and evaluation of drugs for medical use.

The effort to develop a regional cluster in nanoscience has advanced in recent years. In July 2011, Governor Beebe signed a memorandum of understanding with the U.S. Food and Drug Administration to establish a nanotechnology research collaboration between the National Center for Toxicological Research (NCTR) and five Arkansas universities. The MOU provides for creation of a virtual Center of Excellence in Regulatory Science that will include toxicological research associated with nanotechnology and a regulatory science curriculum at the University of Arkansas for Medical Sciences. It also establishes a working group appointed by the Governor and co-chaired by the Director of NCTR and a gubernatorial appointee to coordinate the Center's activities and assist in commercializing its research results.[117] Local business leaders believe NCTR can do for Arkansas what Oak Ridge National Laboratory has done for Tennessee – that is, generate products for commercialization that generate high-knowledge, highly paying jobs.[118]

---

[117] "Beebe, FDA Sign First of its Kind Agreement at NCTR," *Arkansas Business (*August 12, 2011).
[118] In 2001, the Department of Defense deeded 1,500 acres of arsenal land adjacent to the NCTR facility to the Economic Development Alliance of Jefferson County. Envisioning a regional research park, local leaders created the Bioplex on the land, using $200,000 in federal grant money to clear land, and to build roads and utilities. The state has pledges to support the project with tax credits, revenue lands and other incentives. See *Arkansas Business*, "NCTR Has Potential to Create High-Paying Jobs," July 4, 2011.

Arkansas Nano-medicine Center

In January 2012, the University of Arkansas for Medical Sciences opened the Arkansas Nanomedicine Center, which will coordinate statewide nanomedicine research efforts.

Food Processing

As noted earlier, Arkansas is one of the leading agricultural states in the U.S., with about 200 food processing facilities located in the state.[119] Food processing is Arkansas' largest source of manufacturing jobs, accounting for 25 percent of the state's 199,915 total in 2011.[120] The Milken study found that Arkansas enjoyed a "definite comparative advantage" in food processing that, while not in itself a high technology industry, does feature many areas for increased technology and science applications.[121] As Carole Cramer noted in her remarks, Arkansas is promoting research through multi-institutional, cross-disciplinary clusters to promote in-state innovation in agriculture:

- The Arkansas Division of Agriculture supports a cluster that has developed a world class reputation in rice and poultry science, and is now focusing on bioengineering.
- The Arkansas Biosciences Institute leads a cluster of institutions with the NSF EPSCOR P3 Center (Plant-Powered Production) featuring research programs in plant biomass and yield, plant protection, medicine and feed production.

Cramer noted that biotechnology – rather than traditional approaches to crop improvement – is needed to promote innovation in agriculture within the state. She foresaw innovation in value-added and specialty crops and products, agriculture, green materials devised from crops and livestock and veterinary products. She credited Wal-Mart's emphasis on "green" techniques with a major impact on attitudes in the state, and noted that the recent Battelle study identified market opportunities for Arkansas in new food processing and

---

[119] Arkansas is the number one producer of rice in the US and is second in broilers, third in cotton, cottonseed and catfish, fourth in turkeys, fifth in grain sorghum, eighth in chicken eggs and ninth in soybeans. Food processors in the state include Tyson Foods, Frito-Lay, Butterball, Wal-Mart, Riceland, Post and Nestle. See the summary of the presentation by Carole Cramer in the Proceedings chapter of this volume.

[120] Industrial Jobs in Arkansas Declined 1.5 percent Over Last Year," *Manufacturers' News* (October 31, 2011), citing the Arkansas Manufacturers' Register.

[121] US consumers have highly sophisticated and increasing demand for healthy and fresh foods. "Speed-to-market, logistics networks, quality control, and the accurate matching of supply and demand: represent areas for the application of science and technology, potentially giving the state "a competitive advantage in the nation." De Vol, et al., (2004) op. cit., p. 150.

preservation technologies, advanced food packing, and food safety biosensors and rapid food-borne pathogen detection.[122]

Information Technology

In his conference presentation, Jeff Johnson of ClearPointe noted that the overall environment for IT companies, and especially start-ups, is changing, along with the broader IT environment. IT is no longer delivered only by internal resources and on-site staff. Instead, necessary data may as likely come from the Internet or a hosted solution from an application vendor as from internal IT.

This has caused a shift in the IT landscape, he said. The day of the traditional IT provider of software, hardware, and break-fix services is coming to an end. Today's IT companies are more focused on services and how those services are delivered. "We will be more concerned about how data arrives at the desktop or virtual PC than we ever have in the past," said Mr. Johnson. "This shift from on-site IT services to remote delivery has created a host of opportunities for startup companies."

Arkansas had already begun to experience some successes from IT-based startups, he said, including Windstream and Allied Wireless. HP was also bringing a new support center to Conway, Arkansas. "All of these help to build the underlying foundation on which a knowledge-based economy is built," he said.

The Battelle study identified "enterprise systems computing "as one of Arkansas' nine strategic focus areas, pointing out that industries in the state related to enterprise systems employed over 35,000 people in 1,700 establishments. Job growth in this field is expected to grow by more than twice the rate as the average for all Arkansas jobs, and to grow at a faster rate than the national average for computer-related jobs. The University of Arkansas campuses at Little Rock, Fayetteville and Pine Bluff possess core competencies in informatics, sensing and senor networks, and use of information systems to manage supply chain logistics.[123] The University of Arkansas at Little Rock has a unique program in Information Quality that is drawing students from around the world.

---

[122] See the summary of the presentation by Carole Cramer in the Proceedings chapter in this volume.
[123] Battelle study (2009) op. cit., pp. 18-20, ix.

Optics and Photonics

Describing the Battelle study at the conference, Jerry Adams of the Arkansas Research Alliance noted that it identified optics and photonics as one of its nine recommended strategic focus areas. At Arkansas State University, the Arkansas Center for Laser Applications and Science (ArCLAS) operates the largest collection of lasers and support equipment in the United States mid-south region. At the University of Arkansas, Fayetteville, optics research is a key aspect of its physics department and its microelectronics/photonics program. Researchers at the University of Arkansas at Little Rock are focusing on optics in the university's applied science Ph.D. programs in physics. The principal focus of optics research at these university sites is the use of lasers for materials development, processing and manufacturing. Several Arkansas companies are engaged in optics and photonics research and at last four have received federal SBIR or STTR grants.[124]

LEARNING FROM OTHER STATES

Arkansas' leaders often draw on the experience of other U.S. states and localities in developing policies to promote the growth of knowledge-intensive industries. Success of particular policies and programs in other states may not be directly replicable, but some of the principles underlying state innovation policies could be adapted to the Arkansas context. For example, the Arkansas Research Alliance Scholars program is modeled on Georgia's highly successful Georgia Eminent Scholars program, which has been luring "top notch scientists to Georgia's research institutions since 1990."[125] The Arkansas Economic Development Commission "carefully studied the organization and structures, service delivery methods and funding mechanisms of a dozen states to identify best profiles" in the area of workforce development and training.[126] Analysis by the National Governors' Association regarding specific state government and metropolitan development strategies has also been consulted.[127] Speakers at the conference highlighted the recent experience of Arizona, California, and Oklahoma in growing knowledge-based economies.

---

[124] Battelle Study (2009) pp. ix, 23. Invotek, based in Alma, Arkansas, is currently marketing an eye-safe laser pointer.
[125] "Scholars Program Copies Georgia's Model," *Innovate Arkansas* (August 22, 2011).
[126] Governor Mike Beebe's *Strategic Plan* (2009) op. cit. p. 26.
[127] A 2006 study prepared for Accelerate Arkansas at the University of Arkansas at Little Rock contained an extensive survey of state and metropolitan strategies for success in the new economy. Gregory L. Hamilton and Teresa A. McLendon, *Closing the Gap: Ann Examination and Analysis of Per Capita Personal Income in Arkansas* (August 2006) pp. 32-44.

Arizona

At the Conference Dr. William Harris, President and CEO of Science Foundation Arizona, described some of Arizona's state-level initiatives to support research. Dr. Harris noted that he had previously served as founding director general of Science Foundation Ireland (SFI), a $1 billion program for strategic R&D investments in that country, and had served with the European Commission to develop the European Research Council. This European experience, he said, has informed his work on developing Arizona's innovation strategy. On the basis of his European work, he identified five key best practices:

- Invest strategically at the state level in university-industry partnerships.
- Operate with speed and flexibility, and work opportunistically.
- Strive for world-class standards for STEM K-12 performance/education.
- Build partnerships with industry.
- Listen to R&D-driven business entities to support the protection of intellectual property.

California

Although California leads the world in many areas of science and technology, Susan Hackwood, Executive Director of the California Council on Science and Technology (CCST) warned that global changes threaten to erode the state's science, technology and educational infrastructure.[128] Dr. Hackwood cited in her conference remarks a number of "erosion factors" that are destabilizing the so-called "closed-business model" traditionally employed in California — in which targeted R&D leads to targeted product/process development in a discrete organization. The erosion factors are the increased mobility of trained workers, the growth in the research capacity of universities around the world, the diminished U.S. hegemony in markets, and a proliferation of venture capital globally. As a result, "your main competition could be anyone

---

[128] Chartered by the state legislature, CCST is comprised of over 200 of the state's leading science and technology experts. Designed to bridge the gap between "those who know science and technology and those who create and enforce the state's laws and policy," CCST produces reports on the state's scientific and technological activities and supports scientific activity and education. Its sustaining institutions are six state universities, and six national laboratories are sustaining members. Recent analytic work includes studies of California's energy future; use of information technology to integrate genetic/genomic test results to promote personalized healthcare; preparation of elementary school teachers to teach science; nanotechnology in California; and reform of California's STEM education structure. See the summary of the presentation by Susan Hackwood in the Proceedings chapter in this volume.

on the planet and your main market is everywhere on the planet. Your inefficiencies are discreetly outsourced." She noted several specific trends:

- Development of new pharmaceutical products has stagnated during the past decade because "corporate pharma is not conducive to innovation," which often arises from small and more nimble companies.
- The state's workforce retention and workforce competitiveness are being challenged by the competition by other countries and regions for skilled workers.
- State and federal policies, such as United States export controls on technology, tort and labor laws, and over-regulation of industries, can unwittingly retard innovation.
- The quality of California's K-12 education is "very poor," state funding of the university system is declining, and the numbers of science and engineering degrees has leveled off.[129]

Arizona used Ireland's model to create Science Foundation Arizona (SFAz), a public-private partnership jointly funded on a 50-50 basis by the state and industry. SFAz had a mandate to diversify and strengthen the state's economy to enable it to compete on a global basis. Its $100.9 million in funds were committed to support R&D in sectors deemed to be state priorities – wind and solar energy, sustainable mining, personalized medicine, new materials and software supporting the semiconductor industry, and aerospace.[130] In mid-2009, after about two years of SFAz activities, the Battelle group evaluated its return on investment and concluded that it had resulted in 11 spin-off companies, 757 jobs created or retained, 50 patents filed or issued, and 292 scientific publications, and that $2.18 in value had been leveraged for each $1 awarded by SFAz in university grants.[131]

Oklahoma

In his conference presentation, David Thomison of Oklahoma's Innovation to Enterprise (i2E) program said that his organization provides businesses in the state with advice, commercialization services and capital, and

---

[129] See the summary of the presentation by Susan Hackwood in the Proceedings chapter in this volume.

[130] Fifty-six percent of the funds were committed to "strategic research" and another 17 percent to graduate-level research fellowships. See the summary of the presentation by William Harris in the Proceedings chapter in this volume.

[131] Examples of new R&D partnerships generated by SFAz included initiatives in concentrated solar energy and energy storage, mining, pharmaceuticals and the development of jet fuel from algae. See the summary of the presentation by William Harris in the Proceedings chapter in this volume.

serves as a portal to public and private resources.[132] The goal of i2E is to create new, home-grown companies in the state of Oklahoma. "What we want to do," said Mr. Thomison, "is to build from within." Referring to the "three-legged stool" described by Mr. Bendis, he said that i2E focuses on just one of the legs: growing companies within the state. "We are in Oklahoma to help people there," he said, "and we want to leverage our in-state resources to the maximum degree."

To increase the number of successful small firms, Mr. Thomison said, his organization collaborates with universities to provide commercialization services, including assistance in marketing, finance, and competitive strategies. The goal was to teach young businesses how to gain access to capital, good management, and networking.

Mr. Thomison said that his organization also helps Oklahoma firms recruit talent needed to grow their company. In most technology-based start-up companies, he said, the leader and founder is the technician or scientist. "These leaders know the technology and the product, and that's extremely important," he said. "But it takes a team to pull off a commercialization." The Oklahoma Center for the Advancement of Science and Technology (OCAST) has found that it must begin by recruiting a CEO and a vice-president of marketing. As the company approaches commercialization, it must also hire a chief financial officer to position the firm to seek venture capital. This positioning includes demonstrating credible resources and creating a capital plan.

To date, OCAST had helped create 433 client companies. Of those, 140 had raised $359 million in equity funding – 66 percent of it from outside the state – and 44 had received $38 million in grants. For 2009, the impacts on the state included $43.4 million in payroll, $115.6 million in reported revenues, and 251 new jobs. The combined companies had developed 336 new products.

IN CLOSING

As documented in the proceedings of this National Academies symposium, Arkansas' political, academic, and business leaders are seeking to foster greater awareness of the challenges facing the state and are taking a number of steps to foster the development of a knowledge-based economy.

---

[132] I2E website, http://www.i2E.org/about Created in 1987, OCAST is a state government agency responsible for technology-based economic development. As of late 2011 OCAST had helped create 433 client companies, 140 of which had raised $359 million in equity funding (66 percent from outside Oklahoma). The combined companies have developed 336 new products. One OCAST innovation was to hire a "CFO in residence," in effect a CFO capable of serving a number of start-ups simultaneously on a part-time basis in their efforts to secure venture capital. See the summary of the presentation by David Thomison in the Proceedings chapter in this volume

Many of these efforts have been based on commissioned studies that offer frank assessments of the challenges the state faces, *inter alia,* in primary education, financing the formation of new businesses, and securing federal research funding.

      The symposium presented a number of initiatives that are under way to address these challenges. These include initiatives in areas such as nanotechnology research and the manufacture of wind power generation equipment. Arkansas' ability to address these challenges by improving its education, investment and research infrastructure, and by leveraging existing areas of strength to create new knowledge-based companies and jobs will determine the future standard of living and long run economic well-being of its citizens. The proceedings, found in the next chapter, provide detailed summaries of the presentations by the state's business, political, and academic leaders, along with those of senior U.S. government officials and national experts. They highlight the challenges, accomplishments, and opportunities facing Arkansas today.

II

PROCEEDINGS

Session I:
The Global Challenge and the Opportunity for Arkansas

Moderator:
Mary Good
University of Arkansas at Little Rock

Mary Good, a member of the National Academies Board on Science, Technology, and Economic Policy, opened the symposium at the William J. Clinton Presidential Library and welcomed the distinguished participants. She said that the symposium would address the opportunities and challenges of building a vibrant innovation economy in Arkansas.

THE INNOVATION IMPERATIVE: GLOBAL BEST PRACTICES

Charles Wessner
The National Academies

Dr. Wessner began by remarking that while Washington DC has a concentration of policy experts, it is in the regions, states, and cities where policy is implemented and tested. State and local leaders, he added, understand the realities of "locational competition" for jobs, companies, and facilities. Modern communications technology and transport systems mean that businesses have the opportunity to switch to suppliers and manufacturing sites around the world.

To stay ahead in this competition, states and regions need to compete by offering fiscal, cost, and other incentives. Moreover, they must compete on the quality and training of their workforce. In this environment, he said, "we must break away from a pro-business or anti-business dialog" and find out what companies really need to prosper. Universities too must work more closely with industry to understand and meet their workforce needs.

He cited a series of "global mega-challenges" faced by the United States and every other country, including fostering economic growth, developing new sources of energy, addressing climate change, improving and "personalizing" health care, and improving security. "The way we can meet these challenges is by innovating," he said. "The pace of competition is increasing, and we need to innovate through public-private partnerships that bring together our best institutions: businesses, universities, research institutes to cooperate in bringing new ideas to the marketplace. Partnerships are the new vehicles for innovation."

Responses to the Innovation Challenge

Leading nations everywhere are responding to the innovation challenge in similar ways, he said, seeking to provide four essential mechanisms for economic growth:
(1) a sustained, high-level focus on innovation;
(2) consistent support for R&D that leverages public and private funds;
(3) support for innovative small and medium-sized enterprises (SMEs); and
(4) new innovation partnerships that help bring new products and services to market.

He pointed to the example of China, which "does all of these things with enormous focus and commitment," especially by making strong investments in education and training; a strategy to move rapidly up the value chain; effective requirements for training and tech transfer; and making productive use of a critical mass in R&D to generate autonomous sources of innovation and growth. "They are focused, committed, and willing to spend," he said. The United States, by contrast has neglected its infrastructure and, despite the ending of the Cold War nearly two decades ago, neglected to adapt its traditional ways of allocating resources to current realities of global competition. "What we have to do is shake things up a lot," he said, illustrating his point by citing R&D spending trends since 1999.

The U.S. share of global R&D spending, he said, had dropped from 39 percent in 1999 to 34.8 percent in 2010; the shares of Japan and Europe had dropped in similar fashion. China's share, by contrast, had risen from 6 percent in 1999 to 12.2 percent in 2010.[1]

Responses of U.S. Trading Partners

He then described in more detail the innovation strategies of several U.S. trading partners. With a population of 4.5 million, Singapore has the ambitious goal of establishing itself as Southeast Asia's preeminent financial and high-tech hub. The stated task of Singapore's Agency for Science, Technology, and Research (A*STAR), with $5 billion in funding, is to:

---

[1] Battelle, R&D Magazine, December 2009.

- Invest in and attract a skilled R&D workforce;
- Attract major investments in pharmaceuticals and medical technology;
- Invest in public-private partnerships (PPPs), including Biopolis and Fusionopolis ("two of the most advanced S&T parks in the world");[2] and
- Develop new programs to address the early-stage funding challenge for innovative firms.

Even so, he noted that Singapore continued to have difficulty generating local entrepreneurs and the growth of new firms.

Spain, he said, had also adopted an innovation strategy and had moved rapidly in recent years to develop appropriate institutions and policies. It has approved the 6th National Plan for Research, Development, and Innovation (2008-11) with a priority of leveraging R&D and innovation. Its Ingenio ["Genius"] 2010 policy package includes many familiar elements: Public-Private Partnerships for innovation, venture funds, and programs to increase research capacity. This includes more money for R&D, an expanded R&E work force (growth of 7.8 percent per year from 2000 and 2006), and university reforms to increase administration, academics, and financial autonomy.

Canada, too, has developed a formal innovation strategy to improve the business environment by reducing taxes, improving the regulatory environment, and supporting SMEs through an Industry Research Assistance Program. Other features include:
- New programs to support university research
- Research and experimentation tax incentives for businesses
- Attracting star faculty by offering special "Canada chairs"
- Reforming immigration rules to attract and integrate highly-skilled workers and pay them well
- A more direct focus on commercialization through centers of excellence, a Sustainable Development Technology Fund, and efforts to develop innovation clusters around federal laboratories.

Finally, he said, Flanders (a region of Belgium with a population 6 million) has been a pioneer in supporting innovation and commercialization. Its primary strategy is consistent government support for imec the Inter-University Micro-electronics Center, a public-private partnership acknowledged to be one of the top semiconductor research centers in the world. Flanders also provides support for universities, incentives for patenting and commercialization, partnerships to support financing for early-stage technology firms; and sustained

---

[2] Parenthetically, Dr. Wessner noted that Senator Mark Prior of Arkansas had introduced legislation to promote more S&T parks in Arkansas.

outreach to the public to explain the advantages of a knowledge-driven economy.

U.S. Initiatives

While the overall growth in total absolute R&D spending in the U.S. is good news, the downward trend in federal spending as a percent of GDP is less propitious for it is investments in basic research that generate the discoveries that lie behind future innovation. The burden of funding basic research is increasingly falling upon the federal government as U.S. corporations focus more of their R&D dollars on later-stage development. Within this declining federal share of expenditure, Dr. Wessner noted, the Department of Defense, which accounts for more than half of the federal research budget, invests around 90 percent of its R&D funds on weapons systems development, rather than on basic or applied research.

More positively, he added, there is a bipartisan recognition of the importance of R&D for the nation's continued prosperity and security. The America COMPETES Act of 2007, signed into law by President Bush sought "to invest in innovation through research and development, and to improve the competitiveness of the United States." Reiterating this commitment in a major address to members of the National Academy of Sciences, President Obama declared that science and innovation is "more essential for our prosperity, our security, our health, and our environment than it has even been,"[3] and set a goal of raising R&D to 3 percent of GDP, and providing new incentives for private innovation and improvements in math and science education. He also urged a doubling of federal funding for basic research over 10 years at NSF, NIST, and the DoE's Office of Science, as well as new investments in S&T infrastructure, new financing for S&T and innovation, and permanent status for the R&D tax credit for businesses.

Initiatives, promulgated through the American Recovery and Reinvestment Act of 2009 (ARRA) also seek to advance research and commercialization of new renewable energy technologies. The wind energy initiative, for example, extends the tax credit for wind-generated electricity through 2012. It provides $6 billion in loan guarantees for renewable energy projects and transmission projects, grants of up to 30 percent of the cost of building a renewable energy facility, and $11 billion in spending and loan guarantees to advance the "smart grid."

Similarly, ARRA funding is directed toward research on other forms of "clean" energy, including $117 million to expand the development, deployment and use of solar energy in the United States, and $2.4 billion in new grants for advanced battery makers.

---

[3] Presidential address at the National Academies, April 27, 2009.

A History of Government-Industry Collaboration

Dr. Wessner noted that such government support for new technologies is not new for the United States. The federal government has played this role for more than two centuries, he said, citing the following examples:
- In 1798, the government made a grant to Eli Whitney to produce muskets with interchangeable parts, leading to the first machine tool industry.
- In 1842, Samuel Morse received an award to demonstrate the feasibility of the telegraph.
- In 1903, the Wright Brothers fulfilled the terms of an Army contract by demonstrating the first airplane.
- In 1915, the National Advisory Committee for Aeronautics helped the rapid advance of commercial and military aircraft technology.
- In 1919, the Radio Corporation of America (RCA) was founded on the initiative of the U.S. Navy, with a dual commercial and military rationale.
- During the 1940s, 1950s, and 1960s, the federal government was a leader in developing jet aircraft, semiconductors, computers, satellites, and nuclear energy.
- From 1969 through the 1990s, the federal government invested in the forerunners of today's Internet and Global Positioning System (GPS).

"Sometimes we forget how we got where we are," commented Dr. Wessner. "When people say it's really new for Washington to encourage a series of innovative industries, I would argue that the record is compelling in the other direction."

One current strategy of the federal government, he said, is to join with states and regions to promote the formation of innovation clusters. "We think the concept is right," he said, "and that not enough money is being put into it."

He noted that previous Academies' studies have shown that science and technology parks can jump-start the development of innovation clusters by bringing companies into closer collaboration with each other and with a university or federal laboratory. A cluster can also enrich the activities of universities by facilitating joint work with industry. He cited research of Professors Van Looy and Debackere of the Katholieke Universiteit Leuven, who demonstrated that that groups university research teams involved in tech transfer publish more, not less, basic scientific work.[4] "These joint teams are both doing

---

[4] According to Professor Debackere, "We found that groups that collaborate have a reinforcing effect and generate more fundamental scientific output as well as developmental research, as measured in

interesting research," said Dr. Wessner, "and they're teaching their students how to work with industry."

Suboptimal Investments and the Valley of Death

A key challenge for the United States, he said, is how to capitalize on investments in research. A popular myth, he said, is that if an idea is a good one, the market will fund it. The reality is that potential investors have less than perfect knowledge, especially about innovative ideas. He noted that George Akerlof, Michael Spence, and Joseph Stiglitz had received the Nobel Prize in 2001 "for their analyses of markets with asymmetric information;" such information can easily lead markets to make suboptimal investments.

Suboptimal investments, said Dr. Wessner, are also a primary cause of the "Valley of Death," in which many small firms perish for lack of funding before they are able to commercialize their products. During the early stages of developing a product, young firms need access to capital, such as from angel or venture financing. Angel sources are typically quite small, however, and venture capital firms have been moving farther downstream, away from risk. VC investments in 2009 shrank 37 percent from the previous year to $17.7 billion. Only 9 percent of that amount was going into seed-stage deals and 26 percent into early-stage deals.

Three federal programs provide a path across that valley, he said – the Technology Innovation Program (TIP), the Small Business Innovation Research (SBIR) program, and the Manufacturing Extension Partnership (MEP).

Technology Innovation Program (TIP)

The Technology Innovation Program at the National Institute for Standards and Technology seeks to accelerate innovation by supporting high-risk, high-reward research in areas of critical national need. TIP provides funding to universities, small and medium-sized businesses, and consortia for research on promising technologies. Awards are merit-based, with funding through cost-shared research grants, cooperative agreements, or contracts. The

---

number of publications. And industrial R&D feeds academia R&D in providing real problems." See National Research Council, *Innovative Flanders*, C. Wessner, ed., Washington, DC: National Research Council, 2008. In particular, see Van Looy, Bart, K. Debackere, and T. Magerman. 2005. *Assessing Academic Patent Activity: The Case of Flanders.* Leuven: SOOS. See also Van Looy, Bart, Marina Ranga, Julie Callaert, Koenraad Debackere, and Edwin Zimmermann. 2004. "Combining Entrepreneurial and Scientific Performance in Academia: Towards a Compounded and Reciprocal Matthew-effect?" *Research Policy* 33(3):425-441.

impact of the program has been limited, however, because of insufficient funding.

SBIR

The SBIR, initiated in 1982, is a competitive, double-gated innovation system providing merit-based awards to small companies to provide proof of principle and develop prototypes. Phase I awards of up to $100,000 are meant for feasibility and proof of principle research, and Phase II awards of up to $750,000 are to develop prototypes or products that are ready for market or other application. There has been much discussion of a Phase III, for product development and commercialization, but there is no SBIR funding for this.

SBIR awards are financed by a 2.5 percent set-aside from federal agency budgets. The "certification effect" of SBIR funding often attracts private capital and/or increases the chance of winning a public contract. SBIR often provides the first money to help start projects, and may even help academic researchers who have no company. The owners of the Intellectual Property retain control, no repayment is required, and SBIR recipients retain IP. The program was recently evaluated by the National Academies, which reported positive impact on firm formation and growth.[5] SBIR funding has also been used to hire academic consultants and to partner with other firms.

Many states have leveraged the federal SBIR program to boost local growth. For example, North Carolina awards up to $100,000 in matching funds to each company that wins a federal SBIR grant, reinforcing support for high-potential small firms.

Several factors affect a state's success in attracting SBIR awards. The key is that states with more applicants get more SBIR awards. The number of applicants is related to the number of high-tech companies, number of scientists and engineers in the state, state expenditures on R&D, private R&D expenditure in the state, and the number of universities. If an application is rejected, the firm can apply again without prejudice.

MEP

Arkansas can also leverage the federal Manufacturing Extension Partnership (MEP), he said. MEP, part of the Department of Commerce, is a national network of specialists in business and manufacturing that offers help in many forms to small and medium-sized manufacturers. Its 440 centers across the

---

[5] National Research Council, *Early-Stage Capital in the United States: Moving Research Across the Valley of Death and the Role of SBIR,* Washington, DC: The National Academies Press, forthcoming.

U.S. team with industry, as well as state and local organizations, and leverage over $100 million of federal investment into nearly $300 million in benefits to fast-growing businesses.

In conclusion, said Dr. Wessner, "Innovation is the key to how regions and nations compete in the 21st century. It is the key to the growth, prosperity, and security of our nation's states and regions. Resource inputs are essential, but not sufficient. Incentives shape the cooperation required for innovation and this involves institutional change. Innovation policy should not be an afterthought. It is a central mission of government at every level – and our children's future depends on it."

INNOVATION INFRASTRUCTURE AT THE STATE AND REGIONAL LEVEL: SOME SUCCESS STORIES

Richard Bendis
Innovation America

Mr. Bendis, President and CEO of Innovation America, began by commenting on the high level of innovation activity in Arkansas. In defining innovation, he noted that it was not limited to technology. "Innovation," he said, "is the creation and transformation of knowledge into new products, processes, and services that meet market need." Crucial to this process is making the transition from product-based economic development to innovation-based economic development (IBED). "Innovation is not just about products," he said. "It's about ways to do things more effectively."

The goals of innovation, he continued, begin with "intervening at the margins between the public sector and private sector flows of capital." Key steps include addressing this economic transition and capturing the benefits of investments in research and development and in higher education. For every innovative idea or firm, he said, it was essential to reach out to other markets. "When you're working with entrepreneurs," he said, "it's important that they are introduced to the global markets even when there are just one or two people in a firm."

Sector Roles in Innovation

Each sector has a slightly different but essential role in innovation, he said. For the federal government, that role included long-term vision and planning, and the ability to identify gaps and trends in science, technology and innovation. It was also to serve as a catalyst in making strategic investments in under-supported areas and in building partnerships with industry. And it included the use of mechanisms designed to encourage innovation in the private sector.

In academia, he said, the role is more straightforward, focusing on the creation, integration, and transfer of knowledge. The most direct way a university transfers knowledge is through the students who graduate and become innovators. Another way is to develop inventions with commercial potential that become the basis for new firms or may be acquired by existing firms.

He said that within the process of innovation, the role of industry was essentially to create wealth. He quoted Joseph Schumpeter, who wrote in 1942: "The interaction of technological innovation with the competitive marketplace is the fundamental driving force in capitalist industrial progress." [6]

For a region to have its own driving force, he said, requires a "three-legged stool": first, it has to attract companies from other regions; second, it has to retain companies already in the region; and third, it has to create new companies. Where most economies fall short, he said, is in the difficult process of creating new companies. They may also have difficulty attracting companies from other regions, because the only solution to the challenge of small-firm development is to apply "patience, persistence, and consistency."

He turned to the model of public-private partnerships, which drew its effectiveness from the integration of three "inseparable missions": (1) the mission of the university to promote research, public service, and lifelong learning; (2) the mission of industry to create products, processes, and profits; and (3) the mission of government to promote economic benefit, return on investment, and sustainable development for society.

The Effectiveness of Technology Clusters

Some public-private partnerships are situated within technology or business clusters, where innovation may be catalyzed by the proximity and face-to-face opportunities of many actors from diverse sectors. Except for a few notable clusters that have grown and evolved over several decades, such as Silicon Valley, the Rt. 128 community outside Boston, and the Research Triangle Park in North Carolina, the cluster phenomenon has been widely pursued and studied only for a decade or so.[7] Many states have recently attempted to develop their own variation of the cluster model designed to

---

[6] Joseph A. Schumpeter. *Capitalism, Socialism and Democracy*, New York: Harper, 1975 [orig. pub. 1942].

[7] A cluster is defined by Michael E. Porter of Harvard University, a leading student of clusters, as a "geographic concentration of competing and cooperating companies, suppliers, service providers and associated institutions." "Clusters of Innovation," an investigative initiative of Porter and the Council on Competitiveness from 1998 to 2001, developed a framework to "evaluate cluster development and innovative performance at the regional level."

stimulate development, commercialization, and financing of technology-based firms. The concept is a familiar one in Europe – where there is even a "cluster observatory"[8] – and in Asia, where governments have chosen an aggressive role in planning, creating, and funding clusters.[9] In the United States, however, some states have still not done formal analyses of their primary innovation assets or examined the possibilities for clusters that might build on those assets.

The function of clusters, he said, leads to a discussion of the differences between traditional and innovation-based economic development (ED). The competitive bases of traditional ED include such assets as natural resources, transportation facilities, and costs; the competitive bases of innovation-based ED include talent, knowledge, access to the research competencies of both industry and academia, and innovation intermediaries that can connect people to the resources they need. As a generalization, he said, traditional ED is based on physical assets, and innovation-based ED is based on knowledge-related assets.

He stressed the importance of the innovation intermediary, which he defined as "an organization at the center of regional, state, or national efforts to align local assets and resources to work together on advancing innovation." He said that Arkansas had a number of organizations that functioned as innovation intermediaries within the state. "The question is," he said, "how do all of them interact with each other? If you're going to be effective as an innovation intermediary, you have to be able to go all the way from the investigative and technical R&D activities to the market and business analysis of a company. It's rare that one organization has all the capabilities to do that."

The importance of making such large efforts to promote small business, he said, grows out of the vital role of SMEs in the U.S. economy. Innovative businesses have generated 60 to 80 percent of net new jobs annually over the past decade, he said, and employ 30 percent of high-tech workers, such as scientists, engineers, and computer scientists. Similarly, SMEs produce 13 times more patents per employee than large firms, and represent a key source of innovation for large companies with which they often partner.[10]

Most small firms that need assistance are in the proof-of-concept, start-up, or seed funding stage, where they require investments of $500,000 to $2 million to reach the prototyping or similar stage of development. Some angel investors and angel networks do work with firms of this size, and some funding is available from the Small Business Innovation Research (SBIR) program. But

---

[8] The observatory maps clusters in Europe, offers educational resources, and promotes the concept of clusters. It lists 38 different categories of technology and business clusters. http://www.clusterobservatory.eu/.

[9] In addition to the Biopolis and Fusionopolis in Singapore, Asia supports many parks, some of great size. Zhongguancun Science and Technology Zone, in Beijing, supports more than 12,000 high-tech enterprises in seven separate technology parks.

[10] Small Business Association.

venture capitalists do not provide the bulk of small-firm funding. The average VC investment today is $8.3 million, and this funding is generally reserved for firms that already have substantial revenues and even profits.

Proof of Relevance

"We all talk about proof of concept, but you need more than that," said Mr. Bendis. "It's *proof of relevance* that matters in the paradigm today. You have a product people are buying, you can generate cash flow and profit, and the business is scalable. So you have to move farther along to be funded today – beyond just proving you have something that will work."

For innovation intermediaries to be able to help companies move to proof of relevance, he said, they need to understand and be able to explain who is doing the marketing, who is investing directly in the firm, and who is aligning and leveraging its resources. "The innovation model I like to show," he said, "is not linear, but circular. It is related to a life cycle of commercialization from proof of concept to reinvestment of profits back into new companies as they emerge."

He posed the question of why so many SMEs underperform, and listed ten primary factors to consider: "passion, physical and mental strength, self-doubt, belief, foresight, guts, failure, self-discipline, fairness, and integrity." He said that of 150-200 small firms that develop business plans, only about 10 draw the interest of venture capitalists, and only one is actually funded.

At the moment, he said, SMEs had encountered a "perfect storm" of negative economic conditions blocking success. The most important was that 44 states had budget deficits, he said, noting that Arkansas was one of only four or five that had a budget surplus. "You are fortunate," he told his audience. "You can be looking ahead to do things proactively in innovation and entrepreneurial growth."

The first feature of the perfect economic storm was a reduction of angel financing. Angel investors had reduced their activity by 27 percent from the first quarter to the second quarter of 2009, and the amount of investment capital available to angels had decreased by 50 percent in 2009.

Also, he said, venture funding had "moved downstream," as discussed earlier by Dr. Wessner. The average investment by venture firms in 2009 had risen to $8.3 million, and the first quarter of 2009 was the worst quarter in 12-1/2 quarters in terms of total capital invested by venture firms.

A Wider Valley of Death

A result of the widening gap between angel financing and VC financing, he said, was a wider "valley of death" between the start-up stage and commercialization. The financing needs of early firms that had once been about $500,000 to $2 million now extended from $500,000 to $5 million.

Arkansas, he said, had yet to develop strong links with sources of private equity. He showed a graph of VC investments in Arkansas since 1995. A spike in activity showed in 2006, when investments soared from near zero to about $40 million, or 0.15 percent of the U.S. total, but this was followed by a return to near zero in 2007-2008. He said that the state needed to be proactive and "take control of your own destiny," because "unless you find a way to get companies to a stage where they can attract venture capitalists to your state, they're not going to come." He emphasized that the programs being developed in the state were laying a firm foundation at home to support Arkansas' innovation and entrepreneurial ecosystem. "You can't be dependent on the national VC market."

He then turned to best practices in financing. One way to fill the financing gap, he said, is through angel capital – not just individual angel financers, but angel funds. He noted that the Fund for Arkansas' Future was an indication of angel activity in the state. He added that 29 states had angel capital tax credit programs, and that Arkansas had both an R&D tax credit program and an angel capital tax credit, "both of which are extremely important to stimulate early-stage investment." They are also key to creating jobs. During the three to five years after the 1991 and 2000 recessions, he said, nearly all net new jobs in America were created by companies with fewer than 20 employees. "This means we have to focus on small business," he said.

He turned to some innovative entrepreneurial support programs, including "Y Combinator" in Mountain View, California; "Dream it Ventures" in Philadelphia; and the Pipeline program in Kansas, which was a seed funding collaborative work space offering mentors and advisors, donated services, and entrée to funding sources. "This," he said, "is an important type of program to have in the 'have-not' states." The program was founded by Silicon Valley people who realized there was still nurturing to do beyond that successful area in order to widen their net of contacts.

He emphasized the job-creating power of this kind of investment, compared with public investments in job creation. The federal stimulus bill, for example, was projected to create some 4 million jobs, at a cost of $800 billion. He said that the job creation was "going to fall way short," producing about 1 to 1.5 million jobs. Studies estimated that the cost per job – given that not all of the total was intended for job creation – would be about $350,000 to $400,000. In Pennsylvania and Utah, he said, early-stage investment of seed capital was far more effective, with one new job costing $11,000 in Pennsylvania and $29,000 in Utah. "This is something our state legislatures need to be aware of," he said, "that states are good places to invest money."

Innovation-Based Economic Development at the Regional Level

To meet current economic challenges in the United States, he said, innovation-based ED programs had a good record. Ben Franklin Technology

Partners, in Pennsylvania, was the first in the United States, started 27 years ago. The Ohio Third Frontier program, which grew out of the previous Edison program, began about 26 years ago. More recent programs were started by Midwestern states, including the Kansas Technology Enterprise Corporation (KTEC), OCAST in Oklahoma, and the Arkansas Science and Technology Authority.

Some of the attributes that characterized these leading innovation-based ED programs, he said, included longevity, bipartisan support (regardless of which party was in office), independent public-private partnerships, continuous reinvention, private sector involvement, investing for real returns rather than making grants, accountability, and effective leadership.

He showed a diagram of Pennsylvania's industry clusters, which rested on four "pillars of collaboration": innovation, capital, workforce, and support services. With these pillars in place, he said, innovation could proceed from concept to formation to growth to maturity and "reinvention."

Corresponding with these stages were parallel stages of funding: pre-seed, seed, Series A, Series B/C, and Mezzanine.

"What you have in this slide," he said, "is a very broad portfolio of programs through every stage of the life cycle of a company and every funding stage for any technology category. This has evolved over 27 years, so they've reinvented this program every year to determine what gaps they have and what they need to do to fill them."

Another successful program, he said, was Ohio's Third Frontier program. The Third Frontier was going to be the "gold standard" for state programs if the Ohio legislature passed a pending $1.6 billion, ten-year measure to invest in innovation-based ED.

KTEC, in Kansas, was a program Mr. Bendis helped create in 1985 and 1986. It had gone through some financial challenges, he said, like Ben Franklin in Pennsylvania, which had had a 40 percent budget reduction for 2010. "So states are all over the board on their commitment to innovation," he said. "You just have to stay the course."

Job Creation and Economic Growth

Among the most important metrics for small firms were job creation and economic growth. The fastest-growing firms, he said, were known as "gazelles," companies that grow at 20 percent a year for at least five years. Kansas ranked eighth in the nation in the 2008 New Economy Index, which measures gazelles and many other features. "A lot of that has to do with having a committed, sustained initiative for 24 years," he said, "to help support early-stage companies."

He added another success story, the Kansas Bioscience Authority, which he said had been "the most innovation-based economic development program in America for the last 10 years. They wanted to grow the life science industry in Kansas, and that state is not known for life sciences." So they identified the state's strengths in life sciences, and began to look for ways to build on those strengths. They established a baseline tax revenue for bioscience companies and research institutions, measured the actual incremental growth in state bioscience taxes, and directed 95 percent of the incremental growth to the Kansas Bioscience Authority, leaving 5 percent in the general fund. The KBA uses the money to fund programs and repay the bonds.

He said that the second most innovative program he had seen was the Tennessee Technology Development Corporation (TTDC). The TTDC, funded by $120 million in deferred insurance premium tax credits, has used this "new" money to create six private-sector funds headquartered in the state; those funds invest exclusively in potential gazelles and other Tennessee businesses.

He also mentioned the UStar model in Utah, which creates centers of excellence at universities. One incentive is not a carrot but a stick: If the center cannot support itself in three years, its state funding ends and it must reach out to industry for a public-private partnership.

A strategy of some states is to support state innovation councils, as seen in Idaho, Hawaii, Colorado, North Carolina and Iowa. Others use regional economic innovation intermediaries. He noted in particular the activity in Cleveland, Ohio, where Bioenterprise, Jump-Start, Team Neo, and NorTech "all focus on collaborating."[11] Finally, he described a model of "economic gardening," initiated two decades ago in Littleton and Longmont, Colorado, and now adopted by Florida, which has made it the state's primary economic development initiative. This model, based on "growing the economy from within," seeks to identify and find funding for companies that already have products, good growth potential, and fewer than 20 employees. The strategy is to find new sales opportunities and expand on old ones by providing free or low-cost tools and information to small businesses.

He closed by commenting on the abundance of resources already developed in Arkansas, all of which indicated strong leadership and a collaborative spirit. "You have many of the ingredients necessary," he said "to develop one of the leading business ecosystems in the country."

---

[11] For a review of northeast Ohio's regional development programs, see Edward Hill et al., "Economic Shocks and Regional Economic Resilience," in M. Weir, N. Pindus, H. Wial and H. Wolman, eds. *Urban and Regional Policy and Its Effects, vol. 4: Building Resilient Regions.* Washington, DC: Brookings Institution Press, 2012, pages 193-274. See also National Research Council, *Building the Ohio Innovation Economy, Report of a Workshop,* Washington, DC: National Academies Press, forthcoming.

INNOVATION AND COMMERCIALIZATION SUCCESS IN OKLAHOMA

David Thomison
Innovation to Enterprise (i2E)

Innovation to Enterprise (i2E), said Mr. Thomison, was predominantly funded through the Oklahoma Center for the Advancement of Science and Technology (OCAST). i2E had grown into a private, nonprofit company with a staff of 16, he said, attempting to combine the best attributes of public and private efforts. With offices in Oklahoma City and Tulsa, it focuses on technology commercialization services and capital acquisition by helping small firms develop:
- Collaborative agreements with research institutions
- Commercialization services
- Access to capital
- Entrepreneurial development
- Networking opportunities

Its mission is "Home-grown economic development by fostering the birth and nurturing the growth of Oklahoma's advanced technology companies in Oklahoma." "What we want to do," said Mr. Thomison "is to build from within." Referring to the "three-legged stool" described by Mr. Bendis, he said that i2E focuses on just one of the legs: growing companies within the state. "We are in Oklahoma to help people there," he said, "and we want to leverage our in-state resources to the maximum degree."

The goal of i2E, he said, is to create more companies. "It's a statistical game," he said. "Most start-ups fail, so if you can create 200 start-ups instead of 100, you'll produce more successful companies. So we are absolutely driven to create more companies. When you're a flyover state, you have to do it better, smarter, and quicker, and we are a flyover state."

i2E's broad objective, he said, is to reinforce the incremental efforts of companies and support their expansion. "The key to success for a company," he said, "is that it must become sustainable by increasing revenues. If you don't create wealth you don't attract capital."

'Venture Capital is Not Risk Capital'

His view of the valley of death, he said, was that "venture capital is not risk capital; it is expansion capital." Oklahoma, through i2E, had been able to broker about two to six venture capital-funded deals every year for the last six years. He observed that VC firms had become more risk-averse, funding larger deals for firms that had already validated a product in the marketplace. Early-stage firms had to depend more heavily than before on angel capital and state initiatives for proof-of-concept financing.

To increase the number of successful small firms, his organization collaborated with universities to provide commercialization services, including assistance in marketing, finance, and competitive strategies. The goal was to teach young businesses how to gain access to capital, good management, and networking.

Examples of Success

He discussed several examples of successful businesses helped by OCAST. One was a Tulsa company called Access Optics, which had developed a technique of attaching sapphire lenses to optical devices used in surgery. Access Optics had discovered that its devices eventually wore out through repeated sterilization by steam and pressure. The company concluded that the addition of carbon nanotubes to the adhesive would make the product last longer, but could not afford the R&D required. OCAST was able to secure a grant of $150,000 to do the work.

In another case, OCAST helped a faculty member at Oklahoma State University found a successful firm based on his research using infrared light to diagnose prostate cancer. The researcher did not have the technology he needed to develop his idea, and had never worked with the private sector. OCAST linked him with a firm that had expertise in the area, and the firm was able to develop a new subassembly device using fiber optics, nanoparticles, and the professor's technology. Because many other companies were working on prostate cancer, it shifted to pancreatic cancer, and succeeded in adapting the technique. The company has now developed a two-stage plan to sell the device to a larger medical company and then to market its own product independently, adding 15 to 20 more employees in the next two years. "We're not trying to recruit a company with 1,000 employees and get a lot of press," said Mr. Thomison. "We want to grow them 20 employees at a time."

Another i2E activity was to promote commercialization services, and it had formed a company called Seed Step Angels, a member of the American Angel Capital Association. In the previous six months the group had grown from zero to 25 members, and its objective was to convert promising ideas into sustainable commercial entities.

Attracting People Who Want to Develop Inventions

I2E was now attracting more people who wanted to develop inventions. A young doctor from the University of Oklahoma had approached Mr. Thomison with a plan to "build a better tourniquet," for example. A problem for ambulances during a fast, bumpy ride to a hospital is the difficulty in placing an IV in a patient's arm using a traditional tourniquet. He said that because of the problem of finding a vein under those conditions, 9 to 20 percent of people in ambulances who need an IV do not get them. The new device, designed with two sleeves and an opening, forces arteries to "just pop up" to the skin surface and allow easy IV placement. The device is non-invasive, and so does not require FDA approval, and it is inexpensive. "The traditional tourniquet has been around for 200 years," he said, "and we may have found a way to make it obsolete." OCAST is now helping the inventor with a market assessment, a business plan, and communication with ambulance companies.

Mr. Thomison noted that in most technology-based start-up companies, the leader and founder is the technician or scientist. "These leaders know the technology and the product, and that's extremely important," he said. "But it takes a team to pull off a commercialization." OCAST has found that it must begin by recruiting a CEO and a vice-president of marketing. As the company approaches commercialization, it must also hire a chief financial officer to position the firm to seek venture capital. This positioning includes demonstrating credible resources and creating a capital plan. One innovation by OCAST, he said, was to hire a "CFO in residence" who could help several beginning firms, each of which might need financial leadership for only part of each day or week.

To date, OCAST had helped create 433 client companies. Of those, 140 had raised $359 million in equity funding – 66 percent of it from outside the state – and 44 had received $38 million in grants. For 2009, the impacts on the state included $43.4 million in payroll, $115.6 million in reported revenues, and 251 new jobs. The combined companies had developed 336 new products.

"If you're saving people money or increasing their revenues," he said in summary, "people hire. We are importing net wealth. If you're solving someone's technical problem in Oklahoma, or in Arkansas, you're probably solving the same problem for somebody in California. If you're really doing advanced technology, it is a global marketplace."

CALIFORNIA'S INNOVATION CHALLENGES AND OPPORTUNITIES

Susan Hackwood
California Council on Science and Technology

Dr. Hackwood, the executive director of the California Council on Science and Technology (CCST), said she would like to share some recent observations from a state that is known for its science and technology leadership. At the same time, she said that although California leads the world in many areas of S&T, "we are also cognizant of the global changes that are irrevocably changing the creation of innovation and innovation capacity."

She said that Arkansas was clearly demonstrating its emerging strength in S&T innovation, and that there are lessons to be learned from sharing successes and failures. She said she would begin by describing her organization, and move to some of the initiatives in California that are relevant to Arkansas.

The Job of the CCST

She described the CCST as "a unique institution," chartered by the legislature and comprised of over 200 of the State's top S&T experts. "We are modeled in part after the National Research Council and seek to bridge the gap between those who know science and technology and those who create and enforce the state's laws and policy." The Council produces reviews and reports on the state's scientific and technological activities, and generally supports scientific activities, including education. Its sustaining members include the three California public systems of higher education and three private universities,[12] with six national laboratories as affiliate members,[13]. Among current CCST activities are the following:

- S&T Legislative Policy Fellows: This new five-year pilot program, modeled after the AAAS Fellows program, places top S&T students as policy fellows in state legislatures
- California's Energy Future (CEF): The View from 2050: a statewide analysis of issues designed to show the technical potential, costs, and risks of various energy system choices
- Personalized Healthcare Information Technology (pHIT): A project that seeks to demonstrate (1) how information technology may make possible the integration of personalized healthcare data (e.g., genetic/genomic test results) into an electronic health record system, and (2) how medical decision-making can be improved by building a new knowledge-based model for decision support
- A Qualitative Examination of the Preparation of Elementary School Teachers to Teach Science in California: A report that demonstrated how poorly elementary school teachers are prepared to teach science, and how little science is actually taught
- California STEM Learning Network: a blueprint for transforming California's STEM education structure into a 21st-century system where more students are college-bound or workforce ready

---

[12] University of California, California State University, California Community Colleges, Stanford University, University of Southern California, and the California Institute of Technology.
[13] Lawrence Berkeley National Laboratory, Lawrence Livermore National Laboratory, Sandia National Laboratory/California, Stanford Linear Accelerator Center, NASA Ames Research Center, and the NASA Jet Propulsion Laboratory.

- CCST's California Teacher Advisory Council (CalTAC): An advisory group on STEM education
- Nanotechnology in California: A contribution to a report on nanotechnology by the President's Council of Advisors on Science and Technology (PCAST)

'Some of the Indicators are not Good'

She expressed a note of caution, noting that California's innovation economy, despite its long leadership, was now threatened by the "erosion of science, technology, and educational infrastructure." She cited a quotation from Russell Hancock, of the Joint Venture Silicon Valley Networks: "I'm not telling you the sky is falling, but I have a duty to report that some of the indicators are not good."[14] She added, "It's not a done deal, being top dog. Some things are not going too well." To illustrate California's condition further, she read a comment by Thomas Friedman, *New York Times* columnist: "We are the United States of Deferred Maintenance. China is the Peoples' Republic of Deferred Gratification. They save, invest, and build. We spend, borrow, and patch."

In 2000, the CCST released its California Report on the Environment for Science and Technology (CREST), in which the board assessed the status and long-term trends affecting the S&T infrastructure in the state. The most immediate, and lasting, impact of the study, she said, was to raise awareness of the importance of science and technology in the state's economy.

Last fall, her organization took a "fresh look at what's happening," and found that many substantial changes had occurred in 10 years. The consensus of the report was that California was good at "using people and generating ideas," but not so good "at generating people." In other words, California's innovation infrastructure was in jeopardy, and a new assessment was needed for the 21st century. "Whole new economies are emerging at breakneck speeds," she said, "and many people are commenting on the impact on industry and its effects on policy."

One significant need, she said, was for the adoption of a new business model. The 20th-century business plans, she said, were based on a "closed innovation" model, whereby "targeted R&D leads to targeted new product or process development within a discrete organization."

Toward an Open Innovation Model

The traditional closed innovation model, which had provided great successes in the past, was shifting due to five "erosion factors" that include

---

[14] Quotation from the *San Francisco Chronicle*, February 16, 2010.

increasingly mobile trained workers; more capable universities around the world; diminished U.S hegemony in markets; and a proliferation of venture capital worldwide. "Good ideas are widely distributed today; companies need to recognize that not all of the smart people in the world work for them, and that industrial R&D has become a distributed system."[15]

Competition from 'Anyone on the Planet'

In e-business, she said, entrepreneurs today have many opportunities beyond Silicon Valley – "in China, India, and elsewhere." In addition, the structure of competition had changed radically. In 2000, "your main competition was your neighbors, and your main market was your neighbors. Your static website channeled your customers to your phone number."

For e-business today, she said, "Your main competition could be anyone on the planet. And your main market is everyone on the planet. Your inefficiencies are discreetly outsourced. We have moved away from the closed innovation model of the 20th century and are close to an open innovation model. Open innovation is very big, very different, has more players, and things can go in all directions."

A New CCST Study

CCST had therefore initiated a new study, the results of which would be offered to the new governor and legislature of California early in 2011. "A bi-partisan, bi-cameral group of legislators had asked CCST to conduct a comprehensive assessment of California's "science and technology (S&T) innovation ecosystem," Dr. Hackwood said, "analyzing and reporting current global innovation systems and recommending to the legislature actions that should be taken to sustain the state's role as a global leader in science and technology."

The examination and recommendations will take into consideration "the necessary talent, critical components of the entrepreneurial environment, and effective catalyzing of partnerships." Included in the final report will be a look at "the exceptional attributes of the state's federal laboratories, universities, and other unique facilities and networks" as well as the game-changing possibilities on the horizon, such as the better use of technology preparing students for a workforce of varying needs.

This report will engage business and industry, the pre-K-12 schools, colleges and universities (public and for-profit), federal research laboratories,

---

[15] Henry Chesbrough, Haas School of Business, UC Berkeley, CCST meeting, February 2010, http://www.ccst.us/meetings/agendas/2010/2010feb.php

non-profit research institutions, and the public policy sector. It will provide a roadmap to help guide effective investment, avoidance of roadblocks, and support of California's innovation ecosystem.

Four Key Areas of Emphasis

The report was not yet finished, but she offered four key areas of emphasis that had changed significantly in the last decade. In *communications*, she said, "We are the last generation who are not digital natives." The number of Internet hosts had risen from 50 million in 1999 to 700 million in 2009, and the number of users from 360 million to 1.7 billion. Given the new network of networks, she asked, what is the "right metric" for risk? Should California look outward to connect – e.g., to Shanghai? "How should we identify areas of innovation infrastructure and collaboration?"

In *health care*, she said, the new focus is on the personalization of care and on a data management infrastructure. "In health care," she said, "the two most important changes are that the wellness of the human being, not the illness, now takes priority; and second, the amount of data has increased enormously." She said that "the model of big pharma is broken," because it can no longer produce the drugs needed to be successful for something as complex as a brain tumor. "What interacts with what?" she asked. "What do we treat? The complexity of the model is completely different."

Corporate pharma is not conducive to innovation, she said, because innovation is not scalable and "size is the enemy." It costs billions of dollars to bring a new drug to market, "largely because pharma spends so much money on acquisitions. No game-changing drug has come onto the market in the last 10 years." Innovation, she said, arises from small, nimble companies. In addition, health care is being transformed by an explosion of new tools and technologies, she said, and doctors are overwhelmed by large amounts of data. The revolution in hand-held devices, connected to data, is bringing "an extra brain to help them make decisions."

'Other People are Running Faster'

Another area of emphasis, she said, is the growth of *technology innovation* at global scale. Of primary importance are the challenges of workforce retention and workforce competitiveness. "People we have relied on are going back to China. California is not so much falling behind as other people are running faster." She said that as countries around the world now move faster in the knowledge-based economy, regulatory barriers to innovation, such as the International Traffic in Arms Regulations (ITAR), persist in the United States. California-specific issues holding back innovation include poor K-12 education, tort and labor laws, and regulatory control. "We believe in the myth of

California innovation," she said. "California leads, and its innovations are reduced to practice elsewhere. But now we are becoming merely brokers."

The *education* system is another area of focus, she said, because of several negative phenomena. The most familiar is that the quality of K-12 education is "very poor." In addition, the cost of undergraduate college education has risen faster than the rate of inflation, and numbers of S&E degrees have leveled off. She cited a figure from CPEC Fiscal Profiles, 2008, comparing state funding for the university system with state funding for the corrections system. The percentage allocated to the universities dropped from 13.4 percent of the budget in 1967-1968 to 5.9 percent in 2009-2010. The percentage allocated to the corrections system rose over the same period from 4 percent of the budget to 9.7 percent.

In addition, the for-profit colleges are changing higher education's landscape, gaining a fast-growing share of enrollments. She reported that the University of Phoenix had just passed California State University to become the second-largest higher-education system in the country, with 455,600 students as of February 2010.[16]

"These guys are eating our lunch," she said, "but we don't see it. They are providing services students want. What students want is mobile technology. They want to work and learn anywhere, to integrate communication with content and collaboration. To them, social networking is the new learning community, the new community of practice."

She closed by returning to the topic of innovation, which, she said, had become "the latest watch-word." Would innovation become an opportunity, she said, or was it just another "flavor of the month?" The state's leadership position was not serendipitous, she said, but the result of strategic investments in the science and education infrastructure. Without solutions to develop a state budgeting process that sustains this leadership, she concluded, "we risk becoming the sunset state."

EVOLUTION OF INNOVATION IN ARKANSAS

Watt Gregory
Accelerate Arkansas

Mr. Gregory, chair of the Executive Committee of Accelerate Arkansas, offered a brief history of innovation in the state, beginning with the creation in the early 19th Century of the Bowie knife, which he called "Arkansas' first innovation." He said that the knife, known colloquially as the "Arkansas Toothpick," was popularized by Jim Bowie and was innovative

---

[16] *Chronicle of Higher Education,* February 7, 2010.

because of new "technical" qualities: light weight, mobility, large size and ease of use with only one hand.[17]

Much later, and on a much larger scale, came the Wal-Mart chain of stores, which were innovative in quite different ways. The chain began with Sam Walton's first establishment in Rogers, Arkansas in 1962. Several years later it launched an innovation that had much to do with its success – the pioneering practice of keeping track of inventory by computer. By the 1990s, the company had the largest commercial computer database in the United States.

A subsequent innovation was the decision to build its own warehouses so it could buy large quantities of goods at low prices and keep them ready for delivery to its stores on short notice. From this step the company began "just in time" inventory management, building new stores close to their distribution warehouse centers and allowing more rapid restocking as needed. This scheme also reduced the company's significant shipping costs. Today, Wal-Mart is recognized as the world leader in managing supply chain logistics.

Worldwide Growth and Local Benefits

Such innovations have led to worldwide growth – and local benefits. With sales exceeding $405 billion in 2009 and net income of $14.8 billion, Wal-Mart employed more than 2 million "associates" worldwide, 46,000 of them in Arkansas. More than 1,200 suppliers have opened offices in the state since the mid-1990s. In 2009 Wal-Mart spent $15.6 billion for merchandise and services through 1,700+ Arkansas-based suppliers. That spending supported over 62,000 additional jobs in the state, where the company paid $161 million in state and local taxes.

The company, long vilified by various activist and rights groups, has turned to "green" innovations to supplement its proven ability to grow and prosper. In February 2010 it issued a well-publicized pledge – made jointly with the Environmental Defense Fund – to eliminate the equivalent of 20 million metric tons of greenhouse gas emissions from its global supply chain by the end of 2015.

Mr. Gregory described other innovative institutions in Arkansas, some new and others newly repurposed. The National Center for Toxicological Research (NCTR), located near Pine Bluff, Arkansas, was established by

---

[17] Various versions of the Bowie knife, designed in 1830 by Col. James Bowie, had blades ranging from six to 24 inches in length. Bowie gained fame when he used an early version of this lethal innovation at the so-called Sandbar Duel, where he killed three men. In the words of one investigator, "These formidable instruments...are the pride of an Arkansas blood, and got their name of Bowie knives from a conspicuous person of this fiery climate." George William Featherstonhaugh, *Excursion Through The Slave States, From Washington On The Potomac To The Frontier Of Mexico,* 1844.

Presidential Executive Order in 1971 as an FDA research center. In partnership with researchers from other government labs, academia, and industry, it develops, refines, and applies current and emerging technologies to improve safety evaluations of FDA-regulated products.[18]

Innovation as an Economic Development Tool

The Arkansas Science and Technology Authority was created in 1983 by legislative statute as the "first true state-based effort to support scientific and business innovation as an economic development tool." Its mission, he said, was to bring the benefits of science and advanced technology to the state's people. In 2009 it increased its research activities, completing 31 projects totaling approximately $8 million in awards and tax credits. It funded the Arkansas High-Performance Computing Center, a "core resource for the development of competitive research in the state and for economic development benefits." It also supported the Arkansas Research and Education Optical Network, ARE-ON, a high-speed fiber-optic-based Internet communications network linking the state's four-year-public universities.

The Arkansas Capital Corporation Group (ACCG), founded in 1957 as a private, non-profit business development company to contribute to economic development, "today bears little resemblance to its original operations, which focused solely on small business asset-backed loans." Since 1988, the ACCG has led initiatives on entrepreneurship and innovation in all sectors by creating and promoting venture capital funds, SBA lending, multi-state university student business plan competitions, and a statewide Internet initiative known as Connect Arkansas.

The Arkansas Economic Development Commission, also formed in the 1950s, had recently assumed a much larger role that expanded beyond manufacturing to include service and high-technology industries. Today its mission is to create strategies that produce better-paying jobs, support communities, and support workforce training. With an expanded focus on technology-based businesses, it provides special economic and tax incentives to private sector employers that emphasize, among other technologies, clean energy, computer technology, telecommunications, and power grid management.

---

[18] Subsequent to the National Academies 2010 meeting, the FDA and the state of Arkansas in 2011 entered into a Memorandum of Understanding, the first in the nation between the FDA and an entire state, focused on harnessing the relative strengths of the NCTR in toxicological research and the research resources in Arkansas' higher education research institutions, in a collaborative effort, including creating a virtual Center of Excellence in Regulatory Science, to provide a model for the nation that brings industry, academia and government together to solve societal problems.

The Arkansas Biosciences Institute, a research program set up with funds from the Arkansas Tobacco Settlement Proceeds Act of 2000, is a partnership of scientists from universities, hospitals and medical schools. Its mandate is to conduct agricultural research with medical implications, bioengineering research, tobacco-related research, nutritional research, and other related research.

Raising Equity for Small Businesses

A statewide Task Force for the Creation of Knowledge-Based Jobs, appointed by the director of the Arkansas Economic Development Commission in 2001, was followed a year later by Accelerate Arkansas, a volunteer initiative of business, professional and educational leaders in the state focused on increasing the average per capita income in the state through increased emphasis on building a knowledge-based economy. Accelerate Arkansas' first significant act was to commission the most extensive study on the Arkansas economy ever conducted.. The 2004 study, carried out by the Milken Institute and funded by the Winthrop Rockefeller Foundation, recommended actions necessary to raise per capita income in Arkansas (now only about 72 to 75 percent of the national average) to the national average by 2020. At about the same time, the Arkansas Institutional Fund was implemented as a fund-of-funds to invest in equity of private venture capital funds to support the funds' equity and debt capital investments in small technology-based businesses through their first and second rounds of investment. Through September 2008, the AIF had made 7 commitments totaling more than $24 million to VC firms seeking to invest in technology-based Arkansas businesses.

By 2007, Accelerate Arkansas was ready to firm up its strategic plan by focusing on five core strategies that have led to significant state-sponsored initiatives focused on building a knowledge-based economy:
- "Support research that is likely to lead to job creation;
- Develop risk capital for all stages of the business cycle, especially for the funding gap between discovery and commercialization;
- Encourage entrepreneurship and new enterprise development;
- Increase the education levels of Arkansans in science, technology, engineering and mathematics (STEM); and
- Sustain successful existing industry through advancing technology and competitiveness."

Mr. Gregory then turned to Governor Mike Beebe's 2009 Economic Development Plan, which embraced these strategies and placed them in the context of addressing the state's competitiveness at the national and global levels. The Plan includes the following goals:

- Increase the income of Arkansans at a rate faster than the national average
- Expand entrepreneurship that focuses on knowledge-based enterprises
- Prepare Arkansas businesses to compete more effectively in the global marketplace
- Develop economic development policies that meet special needs and take advantage of existing assets in various areas of the state
- Increase the number of workforce members with post-secondary educational training

He concluded by comparing these goals with those of a statewide Task Force for the 21st Century Economy (which was created by the legislature in 2007 and completed its study and recommendations in 2008) that studied the scope of economic development in Arkansas and identified programs and services needed for continued development. He noted that this Task Force's recommendations aligned closely with Governor Beebe's objectives. They began with human resource development, especially STEM education at all levels, and more specifically with workforce education. The Task Force then called for ways to carry innovation into the marketplace, increase support for entrepreneurship, provide additional risk capital, and focus on increased global competitiveness in recruiting businesses and industries. They finally called for an increased emphasis on cyber-infrastructure development and innovation by existing businesses.

In summary, Mr. Gregory stated that innovation efforts in Arkansas have evolved from uncoordinated private enterprise-based efforts, to state involvement and encouragement, to the present day public-private sector partnerships throughout the state that seek to accelerate the conversion of the state's economic resources from traditional economic development activities to those focused on expanding knowledge-based jobs and making globally competitive businesses, thereby increasing the standard of living for all Arkansans.

Session II:
Cluster Opportunities for Arkansas

Moderator:
Paul Suskie
Arkansas Public Service Commission

ARKANSAS AND THE NEW ENERGY ECONOMY

Paul Suskie
Arkansas Public Service Commission

Mr. Suskie introduced his topic by saying that discussions of energy policy tend to ignore the role of the electric utility companies, and he proposed to give that perspective for Arkansas in regard to the new energy economy.

He said that he would address four subtopics: (1) the future energy economy, (2) the historical model of utility regulation, (3) the importance of making a transition to a new energy model, and (4) activities at the Arkansas Public Service Commission. "In particular," he said, "I think 2010 will be a historic year in utility regulation in the state of Arkansas."

He began with one fact that he found "absolutely enlightening." That is, when the National Academy of Engineering was asked for a list of the 20 greatest achievements of the 20th century, it placed the electrification of America in first place.[19] "It's fascinating to look at the remaining 19," he said, "because 14 of them directly require the use of electricity. And of the remaining five, all indirectly use or need electricity. It's an integral part of our economy, our living standards, and our quality of life."

Even today, however, much of the world still lacks reliable electricity. He showed a map depicting areas that have been electrified, and those that have not. In Afghanistan, for example, 94 percent of the country does not have electricity. "What will happen to the supply and demand and cost of conventional generation sources, such as coal, natural gas, and oil, when the rest of the world becomes electrified? If this is not the greatest challenge of our time, it is certainly one of them."

---

[19] George Constable and Bob Somerville, *A Century of Innovation, Twenty Engineering Achievements that Transformed our Lives,* Washington, DC: Joseph Henry Press, 2003.

He offered two quotes about the "utility of the future," the first from McKinsey & Company: "U.S. electrical energy investments of $520 billion would yield energy savings of over $1.2 trillion by 2020 and reduce projected energy use by 23 percent."[20] He commented, "What an incredible opportunity we have. But does our current utility model allow us to promote energy efficiency, and how will we pay for it?"

The second quotation was from the Brattle Group: "The United States will need to spend $1.5 to $2 trillion by 2030 to upgrade its electricity system. To raise and spend capital on this massive scale, the utility industry must represent a sufficiently attractive investment vehicle."[21] In other words, he said, to justify this expense, the utility industry needed to design mechanisms that could ensure the recovery of those investments and promote appropriate economic outcomes.

He commented on the low level of understanding of what public service commissions do, other than control electricity rates. He said that the nation's electrical grid was not only one of the greatest achievements of the 20th century, but also one of the most complex. It was now in urgent need of improvement, he said, in order to incorporate renewable energy sources, improve reliability, and accommodate the continuing rise in demand for electricity.

Changing the Monopoly Model of the Utilities

He turned to the history of electric and gas utilities, which both arose at the turn of the 19th century. Both were inherently monopolies, and the nation needed a way to control monopoly prices. By 1914, 43 states had established regulatory bodies to set utility rates. This had to be done in a way that addressed both the capital-intensive nature of utilities and the need for those who provided the capital to earn a return on their investment. Rates were set to provide this security. For many years, he said, that model worked well. It brought electrification to the country, along with an "incredible lifestyle." The challenge for the 21st century is to adapt that model to new realities.

Those realities include challenges no one imagined a century ago, including global climate change, decreasing supplies of fossil fuel sources, and increasing security issues. The historical model was straightforward in encouraging consumption and selling more units to increase profits, which enabled the utilities to continue grid expansion. This strategy succeeded. A

---

[20] McKinsey and Company, "Energy Efficiency, A Compelling Global Resource," 2010. Access at www.mckinsey.com/~/.../A_Compelling_Global_Resource.ashx
[21] Peter S. Fox-Penner, Marc Chupka, and Robert L. Earle, "Transforming America's Power Industry: The Investment Challenge 2010–2030." Brattle Group Report presented at the Edison Foundation Conference, April 21, 2008.

second strategy was to provide the incentive for capital investment, and this worked as well, with investors responding to the high and steady returns of utilities.

A Regulatory Model that Rewards Efficiency

The problem with a regulatory model that rewards consumption, however, is that it does not reward energy efficiency or conservation. These behaviors can be rewarded, but they require a new model. "How do we keep this historical model," he asked, "that has worked so well and electrified America and provided natural gas to most of it, and now turn it around to promote investments in energy efficiency and renewables?"

The day Mr. Suskie was sworn in as chairman of the public service commission, he said, he offered a quote that he had yet to find wrong in any aspect of his life: "Einstein said you cannot solve significant problems from the level of thinking where the problems were created. If we keep the same regulatory model, we're going to get the same results – counter to renewables, counter to efficiency."

A new model, he said, needs to incentivize energy efficiency, while maintaining our standard of living. It will also have to offer reliable returns on investments in energy efficiency and renewables.

He likened the change needed to the evolution of the telephone. When land lines were dominant, the pricing structure of the telephone monopolies had to be regulated. The deregulation of the telephone business allowed the cell phone to emerge. Regulation of cell phone pricing is not needed because the free market sets prices, allowing the field as a whole to grow rapidly.

The same change is needed in electricity metering, he said, where technology has not changed since the 1930s. In most jurisdictions, customers pay the same rate irrespective of the time of day, season of the year, or wholesale prices. This situation is gradually being replaced by "dynamic pricing" and "smart rates" that allow or incentivize customers to use more power when prices and demand are low. He cited a pilot program in Baltimore, where "99 percent of the customers in the program understood its advantages" and requested to continue it.

Moving Toward the New Model

He said that the Arkansas Public Service Commission had initiated several programs to move the utility industry toward a new model. The first included three steps: (1) In 2007, the APSC initiated a series of consumer incentives; (2) it responded to the historical declining sales of the gas industry through a decoupling mechanism that allowed companies to recover those lost sales; and (3) it opened three dockets in 2008: a transmission docket, to help incorporate renewables into the grid; a search for new ways to increase efficiency; and better use of transmission to reduce generation costs over longer distances.

A second program was designed to provide "more innovative rate making." This included investments in new technologies that brought more efficiencies to electricity usage.

The third, he said, was a "sustainable energy docket" to assess the commission's practices. It planned to issue at the end of the year a guide to best practices, including smart meters, feed-in tariffs, and opportunities for renewables. A "fascinating thing," he said, is that the federal government now requires regional transmission organizations to allow individuals and companies to bid on demand response technologies to reduce market loads. This policy, intended to reduce usage, reverses the historical practice of promoting demand and building new generating capacity.

He ended with a quote from Dr. Steven Chu, the Secretary of Energy: "If I were emperor of the world, I would put the pedal to the metal on energy efficiency and conservation for the next decade." Affirming Dr. Chu's view, he said in conclusion, "The challenge we face is how to change our model to maximize that."

FEDERAL-STATE SYNERGIES

Gilbert Sperling
Office of Energy Efficiency and Renewable Energy (EERE)
U.S. Department of Energy

Dr. Sperling began with greetings from Secretary Chu and from Assistant Secretary of Energy for EERE Cathy Zoi. He elaborated on Chairman Suskie's quote from Secretary Chu, saying "he also talks about efficiency not as the low-hanging fruit, but as the fruit on the ground. He has been going around the country for the last year, pleading with all of us to pick it up."

He said he would begin with an overview of EERE activities. To illustrate the magnitude of the current energy challenge, he said that the total cost of investing in new energy technologies, new sources of energy, and upgrading the grid would be on the order of $4 trillion. This would transform the current energy mix from one that is not sustainable, that creates "energy security nightmares," harms the environment, and "is not affordable," to an energy mix that is affordable, sustainable, more secure, and clean.

He described the activities of the EERE in terms of ten programs, divided into (1) major renewable sources (solar, biomass/biofuels, hydrogen/fuel cells, wind/water power, and geothermal) and (2) major efficiency areas (vehicle technologies, weatherization, building technologies, industrial technologies, and federal energy management). "Our mission," he said, "is to diversify and strengthen sources of energy, increase efficiency and productivity, help make energy affordable, increase our energy security, and address the major concerns we have with carbon and the environment."

TABLE 1 Energy Efficiency and Renewable Energy FY 2009-FY 2011 Budget Table

Programs	Current Approp. FY 2009	Current Recovery FY 2009	Current Approp. FY 2010	Cong. Request FY 2011	$ Change FY 11 v. FY 10	Percent Change
Biomass & Biorefinery R&D	214,245	777,136	220,000	220,000	0	0%
Vehicles Technologies	267,143	2,795,749	311,365	325,302	+13,937	4%
Hydrogen and Fuel Cell Technologies	164,638	42,967	174,000	137,000	-37,000	-21%
Geothermal Technology	43,322	393,106	44,000	55,000	+11,000	25%
Solar Energy	172,414	115,963	247,000	302,398	+55,398	22%
Water Power	39,082	31,667	50,000	40,488	-9,512	-19%
Wind Energy	54,370	106,932	80,000	122,500	+42,500	53%
Buildings Technologies	138,113	319,186	222,000	230,698	+8,698	4%
Federal Energy Management Prog.	22,000	22,388	32,000	42,272	+10,272	32%
Industrial Technologies	88,196	261,501	96,000	100,000	+4,000	4%
Weatherization & Intergovernmental	516,000	11,544,500	270,000	385,000	+115,000	43%
RE-ENERGYSE	0	0	0	50,000	+50,000	N/A
Program Direction	127,620	80,000	140,000	200,008	+60,008	43%
Program Support	18,157	21,890	45,000	87,307	+42,307	94%
Facilities and Infrastructure	76,000	258,920	19,000	57,500	+38,500	203%
Congressional Directed Activities	228,803	0	292,135	0	-292,135	-100%
Use of Prior Year Balances	-13,238	0	0	0	0	N/A
Total, EERE	2,156,865	16,771,907	2,242,500	2,355,473	+112,973	5%

SOURCE: Gilbert Sperling, Presentation at March 8-9, 2010 National Academies Symposium on "Building the Arkansas Innovation Economy."

The EERE budget, he said, normally totals about $2 billion. It increased slightly in FY10 and added about 5 percent in 2011, with a significant shift away from the heavy hydrogen focus of the Bush administration and toward more near-term and applied work, especially efficiency and conservation measures.

The Effect of the Recovery Act

The American Recovery and Reinvestment Act had briefly but dramatically changed the energy budget profile, with the total rising temporarily from $2 billion to $16.8 billion, including "the largest one-time investment in the history of the United States in energy efficiency." Of this $11.5 billion investment in energy efficiency, just over half was returned to the states to spur building weatherization and other efficiency activities. "It's been quite a ride this past year," he said.

Even without the ARRA, which would last only two years, the nation's commitment to energy efficiency and renewables is beginning to increase, and the EERE is a primary source of funding.

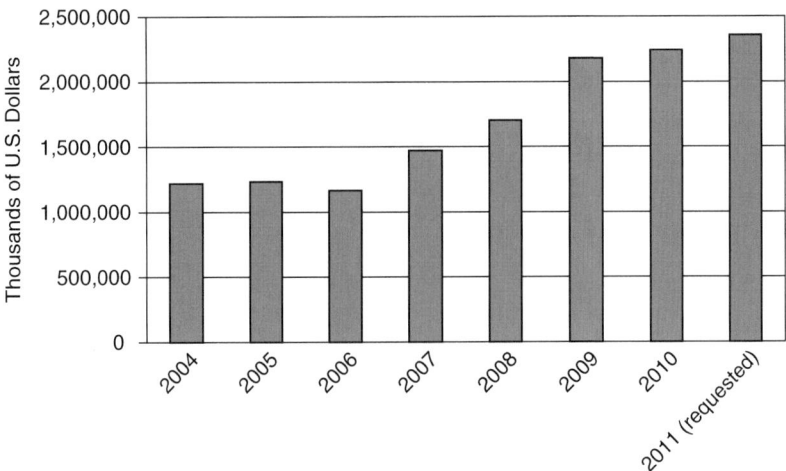

FIGURE 1 Energy Efficiency and Renewable Energy Budget History
SOURCE: Gilbert Sperling, Presentation at March 8-9, 2010 National Academies Symposium on "Building the Arkansas Innovation Economy."

The total EERE budget has risen from about $1.2 billion in 2006 to about $2.3 billion (requested) in 2011.

He discussed the "main planks" of EERE, beginning with how to bring renewable technologies to scale. The goal was to raise the market share of renewables from 1 percent to 30, 40, or 50 percent. As an example, he said that the nation had been supporting about 200,000 comprehensive residential retrofits per year, about 150,000 of them in low-income areas. Some funds were being provided by states, mainly through utility charges. However, there are 130 million residential units in the country, he said, so the challenge was to scale up from 150,000 per year to 5 to 10 million per year, within two years.

To date, he said, only the state of Maine had set a statewide weatherization standard, with a goal by 2030 of upgrading every home in the state and 50 percent of the commercial buildings. California also had aggressive goals established by the California energy committee and the California public utility commission, "but we need more states."

The second plank was "high-impact innovation." He said that when Cathy Zoi arrived as assistant secretary, all EERE programs were reviewed and prioritized according to biggest likely near-term impact. This strategic review was still proceeding, he said, because the current budget is "largely inherited," and substantial changes would not appear until the 2012 budget.

Winning Hearts and Minds

Third was "hearts and minds," the effort to win the support not only of elected officials, but of the American people in order to reach the scale and speed desired. "We have to explain why investments in EERE make sense," he said. He planned to travel to all 50 states, delivering the message that the EERE program creates jobs. "In a few weeks we will bring the 'clean energy road show' here to Arkansas to work with the governor." He said that for each visit, he is accompanied by experts from the private sector on developing the work force, financing, and energy technologies. This group meets with counterparts in business, and local and appointed leaders; the Sierra Club is a prime sponsor. The strategy is to secure commitments from people to have a dialogue about how to move to a clean energy economy. "I have to admit," he said, "that having done eight states, the results have been outstanding. Part of that is due to working with the media and explaining to people what investments in clean energy can do for their economies."

The last plank, he said, was talent. He said that he himself had nearly 30 years of experience as an attorney in developing and financing the capital-intensive energy pipeline and power projects that ranged from renewables to fossils fuels to power plants. He spent years improving the technology development and research process for the oil and natural gas pipeline industry and then creating new strategies to bring that technology to market quickly.

Just before joining DoE, he said, he worked with a start-up renewable energy company. Because the three founders were business people, it differed from "the usual model," he said, in which a technology person is the founder. He worked with the company to develop its financing, from the friends and family stage through angel backing and a "fairly large" series A financing of $15 million. The company was developing hydrokinetic technologies, which use the energy of free-flowing water, such as waves and tides, to provide electricity without building dams.

The Essential Role of Private Capital

He emphasized that new companies in the private sector were essential in transforming the energy economy. There is not enough taxpayer or ratepayer funding to effect change through public funding alone, he said. "We have to leverage private capital. If we don't, we will not succeed in the transformation that is in the vital interest of the people of the United States. So we're embarking on what I call an unprecedented partnership between the federal government, state governments, and primarily the private sector. Having people who understand both technology development and capital formation is absolutely essential."

He returned to the current ARRA investment in EERE, which had been allocated as follows: (1) $3.2 billion to states, cities, and counties as the Energy Efficiency and Conservation Block Grant Program (EECBG); (2) $3.1 billion to expand the State Energy Program, a legacy effort which had existed for 30 years; (3) $5 billion to raise the ongoing Weatherization Assistance Program (WAP) for low-income homes, from its previous funding level of about $200 million; and (4) $300 million for rebates to spur the sale of energy-efficient appliances.

From this package the state of Arkansas received more than $117 million in grant money for energy efficiency and renewables. Another $34 million came as tax incentives to develop wind power, electric vehicles, batteries, and other elements, both to develop the clean economy and to create jobs. "The theory is that investments in clean energy will have multiple paybacks," he said.

Aligning the Goals of Utilities with the Goal of Energy Efficiency

He said that the program placed heavy emphasis on building efficiency because some 50 percent of the electricity used in buildings is wasted. Technology to reduce this waste is available today, but is used only when supportive policies and services are in place, such as national energy efficiency markets that are properly financed and created at full scale. "Those are the most important things we can do in energy efficiency," he said, "as well as removing the incentives for utilities to make more money by selling more energy. Those incentives are misaligned with our national goal of energy efficiency." He praised again Chairman Suskie for advocating the same change in policies.

He turned to specific strategies for increasing efficiency. The first was "recovery through retrofit," or RTR, which grew out of a collaboration of a joint multi-agency/White House task force. Its report of August 2009 identified the primary obstacles to residential home energy efficiency and specified how agencies should put in place by September 2010 innovations in financing, energy efficiency audits, communication with consumers, and work force requirements, all within the authority of existing law.

The next strategy was called R Squared, a new approach to distribution of block grant funding, which totaled some $400 million. He said the "normal temptation of politicians would be to spread that money into all 3400 countries." Instead, Congress agreed with EERE's request to award no more than 20 grants, to do so on a competitive basis, and to require the winners to leverage their grant money by a factor of five. EERE stipulated that the new money could not come from utility rate payer funds or other grant programs, but that they would have to use innovative financing that will continue for the long term. "We're about to make our selections," he said, "and I'm happy to say we're going to stand up a number of models for doing broad-scale residential and commercial retrofits – essentially without any government money."

In transportation, he said, the dominant thrust of EERE is the electrification of the transportation system. The goal is to have a million plug-in hybrid electric vehicles (PHEVs) on the road by 2015. He mentioned the plans of Nissan to build the Leaf, the first all-electric vehicle, in nearby Smyrna, Tennessee.

'We Have to Lead by Example'

The last efficiency program he described was the federal energy management program (FEMP), by which "we have to lead by example." He said that the federal government is the single largest energy user in the world, and an executive order from President Obama compels agencies not only to develop plans to reduce their carbon footprint, but refuses to allow approval of budget requests unless the agency's budget includes sufficient funds to implement that plan. OMB had already worked with cabinet secretaries, he said, "who had great goals but didn't put any money in them, and their budgets were "kicked back."

He then reviewed the major renewables being supported by EERE. In solar energy, he said, the primary goal was to bring down price. The first large-scale solar concentrating plant had come on line in Nevada, and three more were under development. The state energy program money was being used in a variety of locations, such as a 20 MW solar farm in Tennessee. "Our goals are to achieve grid parity," he said, " and to transform solar markets through initiatives that break down market barriers and promote adoption."

The EERE is working closely with the private sector to develop biofuels for transportation, with a goal of 36 billion gallons annually by 2022. Current production is a little over 4 billion gallons, he said, so that "we need a substantial ramp-up in speed and scale. We're looking at not only new feedstocks, and direct conversion of agriculture and wastes into transportation fuels, but we're driving this toward lower cost." For the past four years, he said, the United States had led the world in ethanol production, and about 7 million flexible fuel vehicles were on the road. Also, the department had set a goal for advanced biofuels that reduce GHG emissions up to 80 percent from a 2005 gasoline baseline. He described advances in enzymes and catalysis, engineering of new microorganisms, and novel sustainability indicators.

Potential Contribution of Wind Energy

In the field of wind energy, he said that a report had just been issued that forecast a potential contribution four times larger than the total use of electricity in America.[22] "The challenge is to realize this potential through innovative technologies that reduce cost but, more importantly, deal with intermittent generation and variability, solve the storage problem, and improve durability, cost, and reliability. Some of the technical advances already made, he said, come close to "science fiction," such as the discovery that technologies developed for other uses turned out to reduce friction of the leading edge of the blade.

He recalled his previous work in hydrokinetics, saying, "Hydropower is already a substantial source of power in the United States, but we can both create new technologies that do not have the environmental impediments of a dam and we can repower existing dams with more efficient turbines and devices that have virtually no environmental impact." The U.S. ocean power industry was still in early development stages, but he said that the potential was large, with global ocean thermal energy conversion resources estimated at three to five terawatts.

Finally, he said that the potential opportunities for developing geothermal sources around the country are "tremendous." The EERE had given one grant of a little over $3 million in Arkansas, and throughout the country it had "aggressive goals for cost reduction and demonstration projects." Installed capacity in the United States so far was about 3,000 MW electric, with 132 new projects underway in 12 states. He estimated that enhanced geothermal systems (EGS) had a potential to produce more than 100 GW by 2050, mostly in western states..

He closed by saying that he hoped the overview he had provided would serve to introduce the next few speakers, who would approach several energy issues in greater depth.

---

[22] See Lu, Xi, Michael B. McElroy, and Juha Kiviluoma. 2009. "Global potential for wind-generated electricity." *Proceedings of the National Academy of Sciences of the United States of America* 106(27): 10933-10938. This analysis indicates that "a network of land-based 2.5-megawatt (MW) turbines restricted to non-forested, ice-free, nonurban areas operating at as little as 20% of their rated capacity could supply >40 times current worldwide consumption of electricity, >5 times total global use of energy in all forms."

THE WIND INDUSTRY IN ARKANSAS: AN INNOVATION ECOSYSTEM

Joe Brenner
Nordex USA

Mr. Brenner said he had moved to Arkansas a year earlier, from Nordex in Pennsylvania, where he had been vice-president for production. "I've been struck," he said, "by how diligently Arkansas works to improve its competitive edge in business, industry, research, and innovation." He had come to oversee construction and eventually production at a new state-of-the-art Nordex plant in northeastern Arkansas that would manufacture wind turbines.

He noted that there was much to do before production could begin. Because Nordex was building an entire operation from the ground up, he had spent much of his time for the past five months interviewing job candidates; he had been able to find only 62 out of an estimated work force of 700 to be hired over four or five years. Because of the high skill level they required, Nordex planned to build an academy to train its own people. It already had 300 people constructing the plant, and would also organize a transportation system to distribute its products.

The Wind Industry as an Innovation Ecosystem

He described the wind industry in Arkansas as an innovation ecosystem. "It's the complex relationship between technical breakthroughs that ignite, enable, and depend on more technical breakthroughs within," he said, "and a certain favorable environment. At the end, we need new and better things that improve our lives, the goal of innovation. Arkansas is surely part of that innovation ecosystem."

He emphasized that a "wind rush" had suddenly arrived in the United States, with 35,000 MW of capacity installed in fewer than 10 years. He said that the United States would probably be one of the largest wind energy markets in the world, along with China. "In other words," he said, "national demand for wind energy is driving the manufacturing industry in Jonesboro, Arkansas."

The Nordex plant, he said, would be one of most high-tech and sophisticated facilities on the continent, innovating in two core areas: wind turbine technology and production technology. The end product would be turbines, a tool first developed hundreds of years ago for application in windmills that pump water. These were followed in the late 19th century by turbines that could generate electricity, beginning in 1888 when famed inventor Charles F. Brush built the first wind generator to provide electricity for his own house in Cleveland, Ohio.

Current technology is very advanced, he said, with a full range of sizes for cold, normal, and hot climates; high or low wind speed conditions; and operating adjustments, such as remote monitoring and variable pitch control. He showed a picture of a turbine that would be manufactured in Jonesboro. It was as high as a football field is long; the span of the blades, too, equaled the length of a football field. Such size brought cost efficiencies, he said, so that wind farms now require fewer turbines and less land to produce savings for consumers. Efficiencies gained by such innovations have brought the price of wind energy close to grid parity, he said.

In Arkansas, several conditions favored development of a wind industry, he said. One was its central location in the U.S. market, which is critical for logistics and cost. A second was the available work force, which can be retrained in new skills. To help this retraining process along, the company was developing linkages with Arkansas' educational system, especially with nearby Arkansas State University, and helping design training programs to meet the needs of the industry. A third strength of the state, he said, was the commitment of state and local leaders, who worked together as "true partners" in building the strength of the economy and developing favorable policies for economic growth. "With every new manufacturer that comes," he said, "Nordex wins. The state is building a critical mass of players that can help us attract suppliers. This is the true ecosystem at work."

He said that Arkansas' Governor Mike Beebe had been a major force in building that ecosystem, and predicted that the presence of turbine technology would lead to a boom in wind energy demand and, in turn, spur "fresh thinking" about transmission, grid technology, and the non-technical elements of the ecosystem, including policy, regulation, and investment trends.

At the same time, he noted that "ecosystems can be fragile," and that the best way to support growth was to cultivate demand. He showed a map of the state and the new wind-technology plants in Little Rock and Fort Smith, saying, "It's in our interest and Arkansas' interest to be aligned now because of this map, because of the industry represented here and the economic ripple effect from those jobs."

A Renewable Energy Standard

He acknowledged the effects of the economic slowdown on the industry, including lost jobs and hiring delays, and suggested steps to promote demand. The first was political support for national energy regulation then in Congress—specifically, a national renewable energy standard, or RES, that would require the states to produce up to 20 percent of their electricity from renewable sources by 2020. "An RES," he said, "would represent a national strategy to diversify our domestic energy sources, stabilize electricity prices, create long-term domestic energy supply, and reduce pollution."

He said that even though winds in Arkansas are not as strong as those of some other states, new technology meant that the state can still support wind generation. He said that the average wind speeds in Arkansas are comparable to those in Germany, which produces up to 20 percent of its electricity from wind in some regions. "There are challenges to finding the right locations in some parts of the state," he said, "but siting specialists are quite sure that Arkansas can provide wind energy."

"Arkansas has placed a wise bet on one of the fastest-growing industries in the country and in the world," he said, "and cultivated an innovation ecosystem for wind energy. In the right locations, with the right people and skill sets, and the right leadership, this state will see jobs created, talent developed, and local economies grown."

ARKANSAS' ROLE IN ENERGY TRANSMISSION MANAGEMENT

Nick Brown
Southwest Power Pool

Mr. Brown, president and CEO of Southwest Power Pool (SPP), said that prior to its formation in 1941, none of the electric utilities in that part of the country were interconnected. Connection came about when an aluminum plant in central Arkansas needed more electricity than any utility could supply, prompting 11 regional utilities to pool resources. Today the state's electrical network is highly interconnected and interdependent, and part of the "single largest and most complex machine on earth." The network was maintained after World War II to enhance "reliability and coordination."

The SPP is based in Little Rock, where there are 439 employees, scheduled to expand to 600 by 2012. He said that the operating region of SPP today encompassed 380,000 square miles of service territory, and that "electrically speaking, you can follow the outlets in this room into Canada and Mexico, all of them part of one very large machine." The pool has 56 members operating in nine states. The membership is very diverse, including investor owned utilities, cooperatives, bulk power marketers, and municipal and state agencies.

The primary function of the SPP is to "act as sort of the air traffic controller for the bulk electric network." The pool does not own the lines, but manages the flows of power and serves as a sales agency for wholesale energy across the network. It also acts as a one-stop shop for purchase of transmission service, saving buyers the chore of negotiating separately with as many as 20 different owners as they "cross the footprint."

One of his main concerns, he said, is the capability of the bulk electric network to meet the region's needs. Demand forecasts vary widely; SPP uses the figure of 1.5 percent per year. That sounds like a small amount of growth, he said, but the current 66,000 MW of capacity, provided by 847 plants, would have to increase by about 10,000 MW by 2017 in order to maintain supply for the SPP footprint. This figure could also be achieved by the equivalent in efficiency or demand-side management. "Either way," he said, "this is a big challenge."

He said that what "keeps him awake at night" is the possibility of a "perfect storm" resulting from issues facing the industry today. These include growth in demand; arguments over types of generating capacity that should be used; the difficulty of integrating renewables into the grid; insufficient transmission lines; lengthy permitting for new generation; and aging infrastructure. In three years, he said, more than 50 percent of the existing "fleet" of infrastructure will have exceeded its 40-year planned service life.

Responses to these issues, he said, must be delivered in "shotgun" fashion, with leadership from multiple, overlapping areas. His first recommendation was to maintain a broad portfolio of resources. "It troubles me that people want to take things off the table," he said, "like coal; 40 percent of our footprint comes from coal." He said that the reason for coal's dominance was cost: the price of coal per million BTU on the SPP footprint in 2009 averaged just below $2, while the cost for natural gas exceeded $7. "We need to put things on the table," he said, "not take them off."

Networks Being Asked to do Things They Were Never Designed to do

He also called for expansion of the bulk electric network, which is "being asked to do things today that it was never designed to do, such as processing $3 billion worth of wholesale transactions." Electrically speaking it was the equivalent of the state highway system; "but we need the equivalent of the Interstate highway system to bring more diverse resources to bear on our growing needs."

As an industry, he said, "we spend a woefully inadequate amount on research, development, and demonstration. In comparison with other industries, we should be ashamed. The reason research is cut is the focus on cost, as opposed to value." He singled out the need for more R&D on carbon capture and storage, renewables, and efficiency/demand response technologies.

The 'Saudi Arabia of Wind'

On the need for more renewable sources, he said that the western portion of the SPP footprint had been called the "Saudi Arabia of wind." He said that most of the states in the footprint had an RES mandate, and Oklahoma was currently considering one.[23] As a result, the SPP had received requests to add more than 35,000 MW in generation interconnections for wind power. The pool was currently unable to meet those requirements "without significant expansion in the interconnection network to deliver that wind to load centers."

One of the challenges of wind, he said, is that it "isn't there during times of highest electricity needs. Some say it's not correlated with load; actually, it's perfectly correlated – in a negative direction." Hence the need for a robust transmission network to help take advantage of wind.

His final point, he said, is that "10 percent of our asset base is constraining the other 90 percent of it. If you look at the capital intensive nature of our business, and the rate basis, about 10 percent of that base is transmission, and it is constraining the decisions we're making on the generation and distribution sides of the business." He said that this is the wrong model, and that transmission should no longer be the "last leg in the planning process. It's the tail wagging the dog and we need to change that mindset."

He said that new transmission capacity was being built, but that costs were high and rising. A challenge is the diverse utility pricing zones in the SPP footprint that "don't make sense. What we envision is an extra-high-voltage overlay of the current grid to create the equivalent of the Interstate highway system, linking all our utilities together for more efficiency, and linking it to other regions. This is one conceptual idea of how to deliver wonderful amounts of wind resources from the western portion of the SPP footprint to load centers in the East and potentially in the West as well."

He said that the SPP had identified numerous benefits of expanding the transmission network in and around SPP, including better ability to accommodate fuel diversity; improvements in market liquidity; ability to idle high-cost, high-carbon and high-pollution resources; and increased energy capacity. Unfortunately, he said, the tools for assessing the benefits of transmission are crude, but both quantitative and qualitative indicators did show that "building transmission makes sense." He said that the indicators used so far were very conservative. They also pointed to the need for the higher voltages that allow transmission of "much, much more power over smaller rights of way."

He closed by reiterating that the SPP is "highly interconnected, and highly interdependent. But we are asking it to do a great deal more today that was ever envisioned in terms of meeting our nation's needs efficiently."

---

[23] He also noted a strong correlation between western Oklahoma counties that have lost population in recent decades with counties that have significant wind resources.

Day 2

The State of Technology and Innovation in Arkansas

The Honorable Mike Beebe
Governor of Arkansas

Governor Beebe began by saying that "we are probably in as tough a situation as this country has seen in my lifetime." Notwithstanding the terrible numbers, however, he said that his own state of Arkansas was doing relatively well. "We've lost jobs, more than we've gained – about 24,000 created, 27,000 lost – but compared to other states, it's better." The numbers from across the country were "staggering," he said. "Pretty dismal. Hopefully we're starting to come out of it."

The stimulus helped many states avoid even further budget cuts, he said, "but you have to be careful about governing under such scenarios. Stimulus money is one-time money." When it is gone, those who have used it for ongoing programs will be faced with a dilemma unless there has been a dramatic turnaround, which few people expected.

"Where possible," he said, "I believe it should be used for two purposes. First, for one-time capital projects. Second, for taxpayers, who should get something in exchange for it, because they're paying for it." He offered the examples of a new road, a laboratory, or an energy-saving upgrade – all of which have an actual return that will be long-lasting and beneficial to taxpayers.

"That's where Arkansas stands in so much better shape," he said. "Others may have had no choice than, say, to pay teachers. Fortunately, we didn't have to do that. When our stimulus money is gone, none of those jobs will be affected." He projected that when this recession ended, there would be "an uptick in economic activity for those situated to take advantage of it. And those who are prepared always come out of a recession at a faster growth rate than those not prepared."

Education and Economic Development: 'Inseparable'

He said he had tried to build the state on the two cornerstones of education and economic development. He called them "inseparable in today's world." During the downturn, he said, of the 27,000 jobs lost by the state, many required less education, and many had been off-shored or consolidated. The

23,000 jobs that the state had created were for the most part better-paying jobs requiring higher levels of education and skill.

"That dynamic portends where this country is going," he said. "States that are weathering this recession a little better have recognized that." He said that the state had problems in health care, criminal justice, other areas, but "I argue that if you get the two components of education and economy right, all others are easier to solve. I also argue that you can't have economic development today without education, because you have to have the high-quality workforce. Nothing is more important when companies decide where they're going to start or where they're going to stay."

But even improving the quality of education is not sufficient, he said, without an economic development plan that allows those educated people to find work. Otherwise, "all you are is a farm club for someone else. You've educated people who leave for other states or other countries, in some instances for work."

Arkansas was beginning to get high marks in education. He cited an article from the Tulsa World, entitled "What's going on in Arkansas?", that described excellence in per-pupil funding, test scores, transparency and accountability, standards, and increase in advanced placement students. The one area where the state lags, he said, is per capita baccalaureate degrees, where it ranks 49th in the nation.

"We're going to change that," he said. "There are two main reasons why people don't have a bachelor's degree. One is lack of academic preparedness. The other is lack of sufficient money." The state is trying to address the first lack in several ways. The first is a new systemic pre-K program "that is the envy of the nation now – so the kids don't start behind." This is complemented by more short-term activities with after-school and summer programs for the generation that missed the pre-K preparedness opportunity. Other policies include higher standards, higher expectations, and more advanced placement. The state recently approved a lottery, and 100 percent of its available revenues are targeted for college scholarships. "There will be no excuse for Arkansas to stay 49th in per capita BA degrees," he said.

A Roadmap for Tomorrow

He said that among the state's advantages are its work ethic and entrepreneurial spirit. He noted Arkansas' reputation for successful businesses, beginning with Sam Walton's Wal-Mart, and continuing with Tyson, J. B. Hunt, Stevens, Acxiom, and Alltel, now a part of Verizon. "Those success stories were the basis for what was yesterday," he said, "but they provide us with a roadmap for tomorrow. We're probably not even aware of how our children and grandchildren will live 10, 15, or 20 years from now. But those who embrace technology and innovation and entrepreneurship; make the marriage between education and economic development; and learn that science is the basis for tomorrow's economy will reap the benefits for themselves, their employees, their loved ones, and their region."

He closed by describing the state's use of its tobacco settlement money, which emphasized both research and collaboration among institutions. "I think we're one of only three states left that are devoting 100 percent of their tobacco money to health initiatives. That was the contract with Congress, that the federal government would not take their share out of Medicaid as long as the states spent it on health care. Others have been using it for roads and so on, but my goal was to take as much as I could convince others to take and set it aside for research. Research isn't sexy in the political world, and you may not have much to show for it in the next election. But it's vitally important to our people. With new money, which is what the tobacco settlement was, it was proper and perfect to take a huge portion of that and set it aside. In doing that, we required collaboration among our institutions of higher education. That's the kind of thing that makes a difference – maybe not today, but tomorrow."

Session II:
Cluster Opportunities for Arkansas

Moderator:
Charles Wessner
The National Academies

RESEARCH IN ADVANCED POWER ELECTRONICS: STATUS AND VISION

Alan Mantooth
National Center for Reliable Electric Power Transmission (NCREPT)
University of Arkansas, Fayetteville

Dr. Mantooth, executive director of NCREPT, referred again to the "Greatest achievements of 20th century," published by the National Academy of Engineering, saying that the technology he would discuss was a blend of achievements Number 1, electrification, and umber 5, electronics. This is the point at which grid modernization has to occur, he said, through power electronics, "which is about to take us over the next 20 years and be the new golden era of electronics."

He mentioned that he was educated both in Arkansas, with a bachelor's and master's degree from the University of Arkansas, and outside the state, with a PhD at Georgia Tech. "And I can tell you that our best here are as good as the best anywhere." He said that one of his objectives was to recruit and retain in Arkansas the best intellects to work on this new opportunity in the electric power industry.

The National Center for Reliable Power Transmission, NCREPT, was founded in 2005 as a center for industrially relevant research and education in future energy systems, including power electronics. Main areas of focus include grid reliability, power interface applications, transportation, energy exploration, and geothermal applications. The theme that ties together many of these activities is "extreme environment electronics," pioneered for space and military applications and now being applied for the high-voltage, high-temperature electronics of the grid.

Defining Power Electronics

He defined power electronics as "basically the interface between where we've generated the power and how we want to condition that power specifically for the load." He said that currently about 40 percent of all U.S. energy is used as electricity, but that this figure is growing "dramatically" as we "electrify more of our lives and charge more batteries." Already, more than 30 percent of all electricity generated is processed by power electronics, he said, with a value of some $300 billion. This percentage is projected to grow to more than 80 percent by 2030, reflecting the addition of more variable speed motor drives, computing, environmental controls, and other electrical devices. "The role of power electronics is to manage their operation more efficiently," he said, "not only in electricity generation, but also in industrial, commercial, and residential applications. "

Dr. Mantooth said that an average power electronics system is about 80 percent efficient today, which means that about $60 billion worth of energy is wasted annually. Power electronics can help not by turning the lights down, but by turning them down "intelligently" – that is, when no one is in the room. The more general goal is to manage power flow throughout the grid in the same way we would manage the lighting in a room.

The Logic of DC Current in the House

"Already," he said, "power electronic interfaces are everywhere. That's where we're moving to be efficient and smart. Most people don't realize that the washer and dryer in your home have a power converter between the plug and the motor. It takes the AC, converts it to DC, then it converts it back to DC to turn the motor. If we had DC distribution in our home we could avoid that loss of energy. So one thing our center is working on is new initiatives with industry partners to put not only AC distribution in a home but DC also, with separate plugs. Many of our appliances and computers want to run on DC anyway."

One thing that makes the Arkansas power electronics center unique, he said, is that it puts its expertise to use on grid-related issues. For example, high-temperature power electronics requires materials that will last for 15 or 20 years rather than three. The electric power industry has always been accused of being slow to adopt new technologies, he said, but they are being forward-looking in demanding reliability in the materials they purchase. Laptop computers have a designed lifetime of about three years, he said. "We can't be selling switch gear to the power industry that has a lifetime of three years."

NCREPT began operations in February 2009, with the broad purpose of accelerating advances in technology to use on the grid. What makes NCREPT unique, he said, is its value as a testing and research facility to users across the country and beyond. Currently, it is the only test facility in the world to offer programmability and reconfiguration options at 6 MW. It offers vertically integrated services from basic research through prototyping, testing, and industrial collaboration with companies and universities around the world. It is also establishing close collaborations with the other universities in the Arkansas system.

He stressed the importance of being able to educate Arkansas students in this growing field and then offer them jobs in the state. "This industry already exists," he said. "Electric power, the grid, and companies like Nordex, Mitsubishi, LM Glass Fiber, Arkansas Power Electronics, Baldor, and Caterpillar. This is where these people will go work for $60,000-$100,000 a year to start."

A Test Facility as a Tool for Economic Development

He said that the new test facility was primarily a tool for economic development, a way to move new technologies out of the lab and into field testing so they can be adopted. The first commercial customer was a developer from New York City, who needed to test a 1.6-MW device for a 45-story building. The building had a diesel generator on the roof, and the electronics device was meant to disconnect the building from Con Edison's grid every time the electricity price rose to 27 cents a KwHr. The diesel generator could produce power for 20 cents, and it was used every day to save money on electricity. The device was tested and already deployed.

He showed a picture of a 3-by-5-inch power electronic module designed to drive the motor of a hybrid electric vehicle, such as the Toyota Prius. Current modules require active cooling from the radiator; this more modern version, he said, is able to operate at 250 degrees C, requires no water cooling, and is lighter and more resilient; three of them can drive the car's electric propulsion system. It had won an "R&D 100" award for innovation in 2009. With funding from Rohm Semiconductor, based in Japan, and Sandia National Laboratory, it was manufactured in Fayetteville by NCREPT and Arkansas Power Electronics.

The facility had been selected as an NSF Industry/University Cooperative Research Center in 2009, which currently had 15 industrial members. This center, called the GRid-connected Advanced Power Electronic Systems (GRAPES), is a partnership with the University of South Carolina, and begins with a research budget of about $1 million a year. "That's not a tremendous amount," he said, "but we will grow it. It's the seed." The GRAPES plan to partner with component manufacturers, equipment providers, and electric utilities/ industrial controls companies. Such companies want to be members of the center both for face-to-face contact with customers and to gain early access to students, "our main product." They also have opportunities for shared IP agreements on the research being done at the lab.

He concluded by emphasizing the importance of training. There is a huge manpower gap, he said, because an estimated 50 to 70 percent of the engineering workforce in the power industry will retire in the next 5-10 years, and little hiring was done in the 1980s and 1990s. "We have to get young people into our schools," he said, "and they have to get a BS degree. This power industry needs people at al levels – it's not just about PhDs. This type of consortium may be able to help."

He closed by saying that the Arkansas center is now among the very best schools and programs in the field, including its partners Virginia Tech, Georgia Tech, North Carolina State University, and the University of Wisconsin at Madison. "Power electronics today is very fragmented," he said, "and Arkansas has an opportunity to take a leadership role and create the center of gravity that we need."

REGIONAL INNOVATION CLUSTERS (RIC)

Ginger Lew
National Economic Council
The White House

Ms. Lew began by saying that while Regional Innovation Clusters (RICs) are likely to be familiar in Arkansas, they are less well known at the federal level. The RICs, she said, have the goal of promoting collaboration between the federal government and regions, states, counties, and cities in order to better align resources. In the President's 2011 budget, more than $300 million had been identified to support Regional Innovation Cluster activities at the Economic Development Administration, Small Business Administration, Department of Labor and the Department of Agriculture.

She showed a diagram to illustrate how RICs encourage communities to identify the economic drivers of their regions. "By encouraging collaboration between business leaders, academic leaders, and community leaders," she said, "we want to know how we can link what you're doing to what the federal government is doing."

Pursuing Energy Efficiency Through a Technology Cluster

She described a meeting the previous year with representatives of eight counties and 15 cities in the Pacific Northwest. The group wanted to pursue energy efficiency by forming a technology cluster, and they were trying to apply for federal dollars to support cluster activities. However, they found the application process to be complex and time-consuming. They showed her a diagram of more than 23 different federal program offices managed by six federal agencies, each requiring "an enormous amount of paperwork," some of which was redundant. They were in the second year of pursuing this federal funding, and wanted to know how the federal government could make this process less cumbersome. "That was a strong motivation for the Obama Administration to start looking seriously at these clusters," said Ms. Lew.

"There are those who believe that the national economy is really a collection of more than 100 regional economies," she said, "and by taking steps to promote them, the federal government would be promoting a more vigorous national economy." She noted that universities, including Harvard, and think tanks, such as the Brookings Institution and the Center for American Progress, had recommended that economic development strategies include regional innovation clusters, and urged the Obama Administration to adopt more proactive policies. She added that France, Germany, Brazil, Scandinavian countries, and others already focused on their regional economies in building national strength.

Energy Regional Innovation Centers (E-RIC)

In response, she said, the Administration had begun "this experiment," beginning with a competitive FOA announced by seven collaborating agencies. The purpose of the Energy RIC (E-RIC) was to spur economic development and job creation, as well as research, while accelerating commercial adoption of innovative technologies that increase building efficiency and conservation.

DoE had already begun to develop its Energy Efficiency Hub, funded at $22 million in the first year and up to $25 million per year for four additional years. It will develop systems-based approaches to designing commercial and residential buildings that integrate windows and lighting, natural ventilation and HVAC, thermal inertia, on-site energy generation, and other efficiency technologies. The Department of Commerce and Small Business Administration would provide and coordinate grant funds to encourage the Hub and Regional Clusters. The Education Department, Department of Labor, and NSF would support collaboration between the consortium and recipients of funding under complementary, existing programs.

The goals of these agencies are to promote consortia formation across the region to compete for this combined investment of $130 million. She said that it was "challenging" to get seven agencies to think about doing things differently, and to coordinate agency program requirements, many of which were set by statute. "But at the end we believe it was worth it, and we were able to realize this experiment because of the commitment of senior leaders across these seven key agencies."

Goals of the Pilot Program

She said that the pilot program had several goals:
- Improve energy-efficient building systems design
- Create and retain good jobs
- Shorten the time to award the grants
- Increase regional gross domestic products
- More closely align community college/technical training with regional business needs for a skilled work force.

"We also hope that small businesses and entrepreneurs will get more tailored counseling specific to the types of technologies and businesses that spin out of the cluster. And we hope that innovations from the cluster can be more quickly integrated into businesses with the help of the Manufacturing Extension Program. We believe that the EDA money can promote a vibrant regional economy by supporting the necessary strategic planning, governance, and infrastructure of the cluster." She said that examples of participating entities include the following:

- Under DOE, the national energy labs, universities, and private industry labs;
- Under EDA, state and local governments, universities, regional government coalitions, non-profits, and native American tribes;
- Other stakeholders in the region who might not necessarily be consulted when forming a technology hub, such as neighborhood associations, community-based organizations, labor organizations, venture capitalists, and business councils

Including these stakeholders is essential, she said, "because we see this as an opportunity to increase the well-being of the entire community and region."

"At the end of the day," she said, "our goal is to integrate the effort of the energy hub with sister federal programs so there is a broader benefit – not only for the DoE, but for all the federal, state, county, and local agencies so that we can achieve a multiplier effect."

Regional Innovation Clusters is still a pilot project. "Our goal is to roll out several other pilot projects this fiscal year. They may not be in the energy field. But we believe that we'll take the lessons learned and integrate them into a broader economic agenda and establish the community of practice that can be used not only by federal agencies, but by regional planners as well."

She closed by noting that innovation was not limited to high-tech activities. Two weeks earlier she had met with a group forming its own RICs that spanned the borders of California and Oregon. The region's primary resources were timber and forestry products, industries that have experienced significant contraction and loss of jobs in recent years. While the region has an official unemployment rate of 27.5 percent, she was told that some residents believe it is actually closer to 40 percent.

"These people believe that through smart regional innovation cluster planning, they can work with their core industry of timber, and bring new technologies to reinvigorate their economies. They've explored options such as clean-energy technologies and talked with pellet fuel manufacturers, to help them establish a more diversified industry base. So innovation is also about rural and urban opportunities. And I believe such opportunities are available here in Arkansas."

AGRICULTURE AND FOOD PROCESSING

Carole Cramer
Arkansas Biosciences Institute
Arkansas State University

Dr. Cramer began with the statement that "agriculture is key to Arkansas' economy," and backed her statement with several persuasive facts. The economic impact of agriculture, including processing and distribution, amounts to some $15 billion, providing 268,000 jobs, or more than one in every six jobs held by Arkansans. Overall, the contribution of the agricultural sector as a percentage of GDP was greater than any of the six contiguous states and higher than the national average.

The state is the top rice grower in the United States; it is also ranked second in broilers, third in cotton, cottonseed, and catfish; fourth in turkeys, fifth in grain sorghum, eighth in chicken eggs, and ninth in soybeans.[24]

Manufacturing in Arkansas, R&D Elsewhere

One effect of this agricultural leadership is that some 200 food processors are located in the state, including some of the world's largest: Tyson Foods, Frito-Lay, Butterball, Wal-Mart, Riceland, Post, Nestle, and others. Wal-Mart alone, she said, "brings a cluster of people who want to sell to them." These firms, however, while they do their manufacturing and processing in the state, do most of their R&D elsewhere. Maintaining agricultural leadership, however, requires research and technological innovation to address today's new challenges, which include:

- Agricultural sustainability
- Climate change and its possible impact on crops, pests, disease, and water
- Food safety, including the controversial issue of genetically modified foods
- Nutrition-related health challenges (obesity, diabetes)
- Integrating bio-energy and food production needs
- The 'grow local, eat local' trend

---

[24] According to the U.S. Poultry & Egg Association, "a broiler is a young chicken raised for meat and meat products. Broilers weigh four to five pounds. Broilers are considered mature at 42 to 49 days old." Access at http://www.uspoultry.org/faq/docs/industryFAQ.pdf

Where the Innovations will Come From

Traditional approaches of crop improvement, she said, would not be sufficient to transform economic development in the state. This would have to come from new techniques of biotechnology, but no one yet knows when such benefits will be available. She predicted that innovations, when they come, would be in value-added and specialty crops and products, including health and nutritional benefits, green materials derived from agriculture, new emphases on livestock and veterinary products, and improved aquaculture for species including and beyond the traditional catfish.

In particular, she said, a 2009 Battelle study had highlighted several market opportunities in food processing and safety, including:

- New food processing and preservation technologies. These include (1) advances in infrared surface heating, high-pressure microfluidization, pulsed electric field processing, and cold plasma, and (2) in-line imaging technologies, such as MRI.
- Advanced food packaging, such as (1) a reduced carbon footprint from "green" packaging and more efficient storage and transport methods, and (2) advanced films that have anti-microbial and other qualities.
- Food safety biosensors and rapid food-borne pathogen detection, including (1) hand-held/on-site and portable technologies, (2) molecular diagnostics for high-value protection, and (3) smart films integrated into packaging, including the use of conjugated nanomaterials.

She noted that Wal-Mart, in making "green" techniques a priority throughout its organization and supply chain, had "affected everyone" in the state and set a tone for agriculture as well as industry.

One of the strategies adopted by the state to promote innovation in agriculture, along with other sectors, was to support multi-institutional cross-disciplinary clusters. One of these was supported by the Arkansas Division of Agriculture through its experiment stations and researchers. This cluster had developed a world-class reputation in rice and poultry science. A current emphasis was on bioengineering, including lean manufacturing techniques and nanomaterials.

Another cluster was led by the Arkansas Biosciences Institute, featuring interfaces with both agriculture and medicine, and the NSF EPSCoR P3 Center, or Plant-Powered Production. This P3 program is a collaborative research network of institutions, including 40 faculty members. P3 has several overlapping emphases, including research programs for health, plant biomass and yield, plant protection, medicine, and feed production.

A 'Serial Entrepreneur'

Dr. Kramer described herself as "a serial entrepreneur," offering the example of plant-made pharmaceuticals. She had begun studying a tobacco enzyme that seemed to have promise as a therapy for Gaucher disease, and in 1993 she co-founded CropTech Corp with SBIR funding and ATP grants. The company grew to 42 employees, and in 1999 won a patent for any lysosomal protein expressed in plants. The shock of 9/11 brought the work to a halt, but in 2003 the technology was licensed to Protalix Biotherapeutics. "The valley of death is real," she said, "but a good idea can survive." In December 2009, Protalix completed a deal with Pfizer, which demonstrated faith in the idea that plant cells can be used to make protein-based drugs. Some think these are safer than animal cells now used by biotech companies.

She concluded that Arkansas has many potential opportunities if it learns to combine its traditional strengths in agriculture and food processing with new techniques of biotechnology. The state now has eight start-up companies receiving SBIR support. It has cross-disciplinary strengths in several promising fields, including new packaging, detection, and sensing technologies; vaccines, probiotics, and advanced feeds for poultry and aquaculture; and "green" products, chemicals, biomaterials, and drugs. To make all this happen, she said, required continued public investment in R&D and start-ups, and developing the "great potential for industry consortia."

INFORMATION TECHNOLOGY

Jeff Johnson
ClearPointe

Mr. Johnson, CEO and president, offered a brief description of ClearPointe, some of the barriers it had to overcome, the help the company received to overcome those barriers, and current information technology (IT) opportunities for entrepreneurs in the state of Arkansas.

ClearPointe, he said, is a managed service provider (MSP), which means it is responsible for maintaining a high level of service for customers that depend on an IT infrastructure. The company delivers this service to its target market, which includes companies of 250 to 2000 employees nationwide. It employs 76 IT professionals in its offices in Dallas, San Diego, Northwest Arkansas, and Little Rock, where it's Network Operations Center is located. Since 2005, ClearPointe has grown by at least 30 percent per year, and expects to more than double in 2010, creating 20 to 25 new jobs as it does so.

Mr. Johnson reviewed the challenges to ClearPointe's early growth. These began with access to funding, which he called "the first barrier for any small startup business." Very few new firms have adequate cash to get a new business through the first year, he said, "and we were no exception." At first the company's only source of financing was its receivables. Soon it decided to sound out the venture capital sector, and in early 2002 it was chosen as a presenter at the annual Arkansas Venture Capital Forum. "This gave us access to knowledgeable people who helped us refine our business plan, and monetize our needs so as to meet our financial plans, even though our timing could not have been worse!" The "dot-com" bubble had just burst, and virtually no VC funding was available for IT startups.

The process did lead to a good business plan, however, and the company was able to "boot-strap," or pay as you go, until it was able to attract some angel backing, which in turn led to the interest of local banks. "Bank loans are not the best way to start a company," said Mr. Johnson, "but we had no other options."

The advantage of starting up in this way, he said, was that it allowed – or forced – the firm to prove its model. They were profitable in the first year, which gave credibility to its scaling model.

"All in all," he said, "we put together a lean company focused on profitability, which help sell you to banks and investors. So far, all of our growth has been able to be funded with traditional bank financing, using the strength of the founders as backing. It has also allowed us to keep 100 percent ownership of the company, which will give us more flexibility with fund raising in the future."

The firm was well aware that traditional bank financing can work for only so long. The next big barrier will be to find "cash flow" financing vs. the "asset"-based lending it currently uses. ClearPointe's goal is to reach $35 million in revenue by 2012, which will depend on maintaining current 30-percent year-over-year growth as well as making some strategic acquisitions. It was able to make two acquisitions in 2009 and one in 2010, using traditional financing, but without cash flow lending it will be difficult to fund the mergers and acquisitions schedule the company has set.

Another issue is finding sufficient IT talent, especially in a rural state. At first it tried "growing their own" – hiring new college graduates and putting them through six months of training before placing them in company roles. This proved to work well, but at a high cost for a small company, with a significant number of employees tied up in training for extended periods. Hiring talent from out of state also was difficult, so that the creation of the Engineering and Information Technology College at UALR came as a "godsend."

"We started working with the college more than three years ago, serving on the advisory council." In the past year, Mr. Johnson and others helped the college design a curriculum that would assure ClearPointe of a steady flow of qualified applicants. For example, one of the classes in the IT program has been "Remote Service Oriented Management, A Practical Delivery." ClearPointe itself co-teaches the class, providing its top engineer as both an instructor and author of the course textbook. The company has also been able to hire experienced technicians and managers made available by the changes at Alltel and Acxiom.

The overall environment for IT companies, and especially start-ups, is changing, he said, along with the broader IT environment. IT is no longer delivered only by internal resources and on-site staff. Instead, necessary data may as likely come from the Internet or a hosted solution from an application vendor as from internal IT.

This has caused a shift in the IT landscape, he said. The day of the traditional IT provider of software, hardware, and break-fix services is coming to an end. Today's IT companies are more focused on services and how those services are delivered. The time of the IT "generalists" is over, as they are replaced by subject matter experts who increasingly deliver their knowledge over the Internet. "We will be more concerned about how data arrives at the desktop or virtual PC than we ever have in the past," said Mr. Johnson. "This shift from on-site IT services to remote delivery has created a host of opportunities for startup companies."

Virtually all business-class applications will be delivered over the Web in the future, he continued, which brings great opportunities in helping companies ready their product for hosted delivery. Microsoft is already working to deliver key services such as email over the web much more efficiently than can be done internally. New opportunities include:

1. Hosted applications: Virtually all business-class applications will be delivered over the Web in the future. This creates great opportunities in helping companies ready their product for hosted delivery. Microsoft is already working to deliver key services such as email over the web much more efficiently then can be done internally.[25]
2. Business intelligence: Businesses also need help to better utilize their existing data. Mining current data more effectively supports better business practices. ClearPointe uses mining to isolate issues and trends across all of its clients' networks to help predict outages before they happen.
3. Data center opportunities: Another opportunity is to provide secure, reliable data center space. This has not been available locally, he said, and is one of the largest hurdles ClearPointe has had to overcome.
4. Security services: These services have expanded from watching for intruders to minimize exposure into developing broad strategies for total prevention of loss of data.

---

[25] Many hosted applications have evolved into a type of cloud computing. See, for example, Bussey, J. (2012) The sun shines on 'the cloud'. *The Wall Street Journal*, July 13: B1.

5. Remote management: By watching over how all these delivery methods work together, a firm such as ClearPointe can help lower the total cost of IT management, and in some cases help re-allocate resources or raise productivity. Remote management of IT infrastructure is one of the fastest growing segments of IT today.

Arkansas had already begun to experience some successes from IT-based startups, he said, including Windstream and Allied Wireless. HP was also bringing a new support center to Conway. "All of these help to build the underlying foundation on which a knowledge-based economy is built," he said.

He concluded that the opportunities for entrepreneurs and startups in Arkansas "are really pretty good." He noted that the lack of high-level competition for IT resources, common to states like California or Washington, may be an advantage. "We also have the luxury of being able to research new technology and start-ups in other states to discover which ones are succeeding, and which ones may fit our situation."

Access to funding remains the highest hurdle to overcome, he said, but financial conditions may be helped by some proposed changes in SBA lending. However, that second or third stage of financing and financing for acquisition continue to be difficult for small firms.

In summary, he said that the state's potential to build a successful IT industry had improved considerably. "With programs like the EIT College at UALR," he said, "hiring the right people to fuel our growth has become less of a problem. And the changes in how IT will be delivered in the future mean that the opportunities for new startups in Arkansas are tremendous."

NANOTECHNOLOGY

Greg Salamo and Alex Biris
University of Arkansa, Fayetteville
University of Arkansas at Little Rock

Dr. Salamo was the primary speaker for Drs. Salamo and Biris, who collaborate in nanotechnology research. He opened by defining nanoscience as "the effort to understand and design structures at the nano size[26] and seek their application." The Arkansas effort in nanoscience is a collaboration among partner institutions throughout the state university system.

---

[26] He defined "nano size" for his audience in terms of atoms (100 atoms in a line would equal about 10 nanometers) and a human hair (the diameter of a hair divided by 100,000 would approximate one nm).

He began by addressing the question, Why are nanomaterials the driver of innovation? "The potential for nano," he said, "lies in the scale of the materials. This is a new way of looking at them. As you make something small, its optical, electrical, mechanical, and other properties change. They may change so much that it's like having a new material. That's why it's so exciting; that's why we love what we do." As one example, he illustrated how a material that flows easily at large sizes will flow with great reluctance as it approaches the nanostate.[27]

He said that new materials have inspired innovation throughout history, and that "we make the best nanoscale material in the country." He said that state-of-the-art nanoscale imaging tools "allow us to see single atoms, and that this changes the ball game." His group is a collaboration of both experimentalists and theorists.

The Potential of Nanomaterials

He cited a series of examples where nanomaterials have the potential to create new technologies in health care, energy efficiency, and renewable energy. In health care, he said, nanotechnology may contribute to cancer diagnosis through the ability to detect single cancer cells flowing in the bloodstream. "We can put something inside that cell and then hit it with laser light that causes it to explode," said Dr. Salamo. He mentioned several other examples, including Dr. Biris' work on the use of carbon nanotubes (CNTs) to label cancer cells in the lymph, blood, and tissues of live animals and even to follow the movement of a single tagged cancer cell in the ear of a rat.[28] He also described a technique of coating cancer cells with graphitic coated magnetic nanoparticles and heating them with radio frequency waves, which results in the death of 98 percent of the cancerous cells after 20 minutes or less of treatment.

He and his colleagues have also reported several techniques of using nanoparticles to increase energy efficiency. For example, he described a nano-bio material that showed ability to reduce friction in mechanical systems, thus saving energy. He also reported on nanoscale oxide materials that can convert waste heat to electricity. This may be applied in automobiles, he suggested, which are only about 30 percent efficient in converting energy to the work of driving; new materials could be used to absorb and use that waste heat. A third area under research is the use of ferroelectric quantum dots as memory elements that are 10,000 times smaller than current memory materials. Finally, he described a new solar cell nanomaterial that can gather solar energy more

---

[27] There are many such examples. At nanoscale, opaque substances may become transparent (copper); stable materials combustible (aluminum); and insoluble materials soluble (gold).
[28] Alexandru S. Biris et al, "*In vivo* Raman flow cytometry for real-time detection of carbon nanotube kinetics in lymph, blood, and tissues," *J. Biomed Opt,* Vol. 14, Issue 2, 2009.

efficiently than current materials by absorbing a larger portion of the solar spectrum.

Such work at the University of Arkansas, he said, had attracted NSF support for a Materials Research Science and Engineering Center (MRSEC) and resulted in six spin-off companies with more than 40 employees working in a state-of-the-art fabrication facility. In 2009, MRSEC authors were credited with about 150 publications that received more than 3000 citations.

More Speculative Ideas

Beyond this already-published material, he said, his group was engaged in a series of more speculative but exciting ideas. One is to place nanoparticles in a certain order so as to carry electricity with high efficiency. Another electronics project is the exploration of a strong chemical bond between copper and manganese that may lead to a new superconductor. This work is based on the observation that atoms arranged in single layers behave as a new material. Another phenomenon he is studying is the use of particulate grafting materials that have stimulated bone regeneration in 43 human pre-clinical cases and 36 goats. Finally, he reported the use of nanostructural titanium dioxide nanotubes to coat tissue implants in the body so as to enhance tissue regeneration.

He closed by noting a consequence of all this activity: The Arkansas university system, he said, now leads the nation in supplying nanomaterials to research organizations across the country.

Session III:
Federal and State Programs and Synergies

Moderator:
Barry Johnson
Economic Development Administration
Department of Commerce

THE ROLE OF THE ECONOMIC DEVELOPMENT ADMINISTRATION

Barry Johnson
Economic Development Administration
U.S. Department of Commerce

Mr. Johnson introduced his panel by reviewing the current status and mission of the Economic Development Administration (EDA). He emphasized that while federal programs are coordinated from the nation's capital, they were only meaningful in the local context. He said that the EDA had been eager to support the states and regions, especially by promoting more opportunities to innovate. Part of the EDA's current effort, he said, was "Innovate EDA," a strategy to improve how activities are performed and sharpen the agency's mind-set on interacting in regional development.

The EDA's investments should go, he said, to places that have demonstrated a commitment to regional collaboration. "It's not just about building a road or a bridge that puts people to work," he said, "but tying the goal to a greater plan for regional prosperity." Such plans, he said, include funding research, investing in tools that regions can use, asset mapping, and investing in alliances that promote regional collaboration.

He also said that the department was about to announce the Technology Commercialization Enterprise Development Alliance, a group of business accelerators like "incubators without walls." These would provide resources to help firms move from an idea to commercialization and help them locate the funding they need to cross the valley of death.

In addition, he described plans to staff the EDA's regional offices with people dedicated to Regional Innovation Clusters (RICs). "As more of our constituents and partners shift to this way of doing things, you'll have regional resources as well as a national resource to help."

He said that EDA believes that its "policies have to change, because the world has changed. When you have unprecedented change, you need new responses. So we are going from silos to collaboration. We believe that this has to happen at the regional level" as well as the federal level.

Defining RICs

He offered a formal definition of RICs as "geographically bounded, active networks of similar, synergistic or complementary organizations that leverage their region's unique competitive strengths to create jobs and broader prosperity." He said that, on average, jobs within clusters pay higher wages, and regional industries based on inherent place-based advantages are less susceptible to off-shoring. Because RICs are locally led, they are able to stabilize communities in various ways: by re-purposing idle manufacturing assets, engaging underutilized human capital, contributing to improvements in the quality of life.

He said that five key components should be considered when defining unique regional assets:

(1) "the economic base: what you make, including your existing and prospective industry clusters;
(2) talent: workforce skills and the human capital base;
(3) particular local conditions: location, infrastructure, amenities, factor costs, natural resources;
(4) innovation and ideas: your capacity to innovate and generate new ideas; and
(5) entrepreneurship: your capacity to create companies wholly new or from existing firms."

Arkansas' Assets

Arkansas itself had many of the assets that support RICs, he said, including the following:

- "A strong network of higher-education institutions,
- Geography and infrastructure that allow for efficient transport of goods and services,
- Good health-care facilities,
- Low factor costs for starting and operating a business,
- A diverse mix of industries in the state's economic base, and
- An integrated network of local, regional, and state development organizations."

RICs are diverse, he said, and found throughout the country; therefore, there could be no standard definition. They support a wide array of industries, and vary in size, shape and reach. They often cross local, county, and state boundaries, and may be urban or rural.

For the energy RIC, or E-RIC, he said, a federal collaboration would grant the funding opportunity to a consortium. He disputed the notion that all boats must rise in a region for a RIC to be successful. "I believe that there is one boat in a region, and it *must* rise. Everyone who is not contributing is pulling away. Leaving certain organizations behind is a costly thing. Who in your ecosystem is not at the table that should be at the table? They all have some role to play."

He summarized recent EDA activities in Arkansas. In 2009, these included 14 investments to support planning and implementation efforts aimed at encouraging regionalism and clusters across the state. A recent EDA Technical Assistance Grant helped establish the Center for Regional Innovation at the University of Arkansas at Little Rock, and a $1.75 million public works grant to Arkansas State University at Jonesboro helped establish the Arkansas State Biosciences institute Commercial Innovation Center.

He concluded by commenting on a $2 million EDA award that had recently been approved to convert the old Rock Island Railroad bridge into a pedestrian and bicycle crossing to link the river market areas of Little Rock and North Little Rock, completing the Arkansas River Trail. "We can't grant opportunities unless they are led regionally," he said, "so I know that some of you made this happen."

DISCUSSION

Dr. Good suggested that the state had two pressing needs. "One is to improve the access to very early stage capital for startup firms. We're beginning to get a fairly decent deal flow in the state, but not enough to take care of the ones we would really like to. Second, we have not been able to find enough funding for some of the innovative things we've done in the state, even though our budget is just a rounding error for the DoE. We are trying to bring others to the table who do have capital, and one conversation is with a network of foundations that want to target their giving at economic development opportunities. This is high among our priorities."

Another participant said that in a state employment bill is a provision to renew a program called Build America Bonds that was a popular part of the ARRA. The bonding authority is dispersed to the states, which in turn disperse it to counties. The federal government pays a portion of the interest payments on the capital. "One of the bond programs, the Zone Bond Program, may be used for some of the early stage funding your talking about. This is a little-known provision, something to explore."

INITIATIVES OF THE MANUFACTURING EXTENSION PROGRAM

Roger Kilmer
Manufacturing Extension Partnership
Department of Commerce

Mr. Kilmer, director of the Manufacturing Extension Partnership (MEP), said that a current objective is to strengthen the "linkages back into the innovation and entrepreneur side of things." MEP is a network of centers in all 50 states, and is "very much a partnership organization, both at national and local levels." It has a staff about 1600, who are located in 440 service locations and 60 MEP Centers through the regions and states. The MEP reaches some 32,000 manufacturing firms and completes 8,000 projects per year.

"What we do," he said, "is really based on what the manufacturers need. We help them find a solution. For example, we partner with community colleges and universities in training human resources, and in many cases we are located at universities to strengthen the linkage between the human resource side and the small to medium-sized manufacturers."

A Portal to Connect with Solutions

He said that in Arkansas, the best portal to connect with the resources of the program is Arkansas Manufacturing Solutions (AMS). This program focuses on how to scale up small firms by working from a national perspective. It focuses more on helping existing manufacturers than on creating new companies, an activity he called "economic gardening." He suggested further that there is or should be a marriage between existing firms and new ones. "If we could find a better way to bring them together," he said, "we could develop more economies for both sides in getting new technology into commercial products and processes."

The traditional emphasis of MEP, he said, is to focus on listening to manufacturers' short-term needs, as opposed to trying to push actions "from our own toolbox." MEP Center projects include:

- Business growth services: what is best for that business in its particular industry and situation,
- Technology services to develop products and processes,
- "Lean" techniques that encourage continuous improvement,
- Quality systems and other standards,
- Advice on energy, environment, and sustainable services, and
- Talent development to meet future manufacturing needs.

Nationwide Impacts

He said that the MEP works hard to quantify what it does, beginning with an independent, third-party survey of manufacturing clients. The most recent client data, he said showed significant nationwide impacts, including $3.6 billion in new sales, $5.5 billion in retained sales, $1.7 billion in capital investment, and 52,948 jobs created or retained (in FY2008). He said that the jobs figure was especially gratifying.

AMS services had also helped clients obtain good results in Arkansas, he said, including $592 million in new and retained sales, $25 million in capital investment, $12.7 million in cost savings, and 3,335 jobs retained and created.

Because manufacturing techniques and challenges are changing so rapidly today, the MEP is having to move beyond its historical focus on productivity. "We've been looking at the environment manufacturers operate in, and trying to focus on new opportunities." He began with the effects of globalization, beginning with outsourcing. "You always worried about the competitors down the road, and now you're having to do the same thing but on an international scale. Supply chains are a fact of life," he said. "Two-thirds or three-quarters of small manufacturers are supplying to some other company, so understanding that is key. Companies need to be innovative, and they need to be more competitive. It's products that bring the home run, but today it's also processes and services and bundling them together. We have to look at different business models, where instead of trying to do everything yourself, you are really partnering to provide a complete set of products and services."

Technology is a principal driver for this, he said. When the MEP was created in 1988, its sole mission was to transfer technology out of the NIST labs to small and medium-sized manufacturers. "We learned early that the kind of technology we were rolling out was not what the small guys needed. So we changed, and said let's look more closely at market pull: What do those manufacturers need, and how can we provide access to that?" Now, some 21 years later, the MEP is at a point where new technology is vital to improving the manufacturing, and the manufacturers are ready for it.

Building Sustainability into Design

Another new concern for the MEP, he said, is sustainability and green techniques of manufacturing. "We're not talking about your father the tree hugger; we're looking at it from a business perspective and asking what are the right things for a manufacturer to do. That comes from both the sales side – there is a market for green things – and also from an operations view: how do you build sustainability into the design, production, and processes? Can I use cleaner chemicals while still trying to conserve energy? At the end of the process can I bring that chemical back and use it somewhere else in the process? Sustainability is the complete 'life-cycle-plus' that includes all of those things."

From the perspective of the MEP, he said, "we've gone from productivity to how do we get companies to focus on growth. And how do we bring technology into the equation." One challenge, he said, is to bring a new mindset to a company – to get them thinking positively about how they can begin to grow the company. After years of difficulty for companies who want to invest in necessary change, he said, "the good news is that now we've got their attention. They know they need to adapt to global competition." He defined MEP growth services as "providing a reliable scientific system that guides companies through the creation of new ideas, discovery of market opportunities, and the tools to drive the ideas into development." It addresses every input to profitable manufacturing growth, including technology acceleration, supplier development, workforce improvement, sustainability, and continuous improvement.

Adapting to Global Competition

There are three strategies for doing this, he said. One is to help companies find, filter, and fast-track ideas. An example was Eureka!, an MEP program to help companies go through a structured analysis to develop the ideas they already have, identify the best opportunities, and develop a plan to address them. A case study was that of a Wyoming company called Precision Analysis, which makes hot water testing kits. With MEP advice, the company was able to re-vamp its marketing message, create a new home water testing kit in line with EPA regulations, develop and release the new product in five months, and double sales within five months.

Another example was a partnership with the Department of Commerce and others called ExporTech, a service to help companies enter and expand in global markets. For example, ExporTech helped the Wilco Machine and Fabrication company of Oklahoma visit the Middle East and establish relationships, negotiate a joint venture in Brazil, and raise exports from less than 8 percent of total revenue in 2008 to 51 percent halfway through 2009.

Finally, MEP has an emphasis on diversification into new customers and markets. An example from the MEP Center in Michigan, where auto suppliers are searching for new partners, is the J.C. Gibbons Manufacturing of Livonia. MEP helped the company move from automotive to medical appliances while using the same equipment and same manufacturing processes while retaining 25 jobs.

Accelerating Technology

Another strategy of MEP is technology acceleration – "systematically identifying and capitalizing on opportunities to leverage technology into the process, products, and services of manufacturers." The challenge, he said, was to link processes to the manufacturer and ultimately the market. "That's the key to this," he said. "We've been swimming upstream, fighting these battles you all know about, and now what we've done is focus the MEP Centers to be that connector. We are not the technology experts. I have an engineering degree and work in a scientific institute, but the technology is so broad and changing so fast there's no way I can keep up, or my Centers can keep up. So the linkage back to the experts is still a key part of solving this."

The issue, he said, is one of connection and scalability. For the manufacturers, there are many links – to their customers, either in an OEM supply chain or another kind of vendor relationship; to the technology sources; to capital and investors. "The question is how do you pull all these components together in a way that can get to efficiency and scalability."

One approach of the MEP, which is at the pilot stage, is the National Innovation Marketplace, an online approach to connect the different links in the value chain. Through the Arkansas Manufacturing Services, MEP had created an Arkansas-specific portal into the National Innovation Marketplace that lists available technologies and opportunities for companies. This is called the Arkansas Innovation Marketplace (AIM), which seeks to provide a window into all the intellectual property and requests and capabilities of its entrepreneurs, inventors, and companies in the state. To date, AIM has about 50 technologies posted and 50 company "needs and wishes."

Finally, he described the MEP sustainability initiative, which seeks to help manufacturers gain a competitive edge and to maintain profitability and job creation while increasing energy efficiency and reducing environmental impacts. In partnership with EPA and the Green Suppliers Network, MEP has for six or seven years helped companies identify where they can make savings and move from a compliance mentality to a search for constructive changes on the front end that can avoid costs and add to the bottom line. "We work with the waste and water management folks and suppliers to the utilities to focus on less waste, less energy usage, more efficiency, and better business practices," he concluded "That also feeds into the concept of clusters in regions."

UNIVERSITY-INDUSTRY PARTNERSHIPS

Marc Stanley
National Institute of Standards and Technology (NIST)

Mr. Stanley, acting deputy director of NIST, expressed his enthusiasm for "trying to grow new companies that have difficulty finding early-stage investment money." He also presented a picture of recent trends that have not been favorable to the development of technology-based business development.

He began with two sets of slides that illustrated what he called "disturbing trends" in R&D investment. The first set (Figure 2) began with a chart of R&D intensity, where the United States lags eight other countries.[29] A second chart (Figure 3) showed that federal spending on R&D as a percentage of GDP has declined steadily since the mid-1960s.[30]

The second set of slides related more particularly to industry's relationship to R&D. The first slide showed an accelerating divergence since 1953 of industry spending on basic research, which has been remained nearly flat, and industry spending on development, which has risen rapidly in the past two decades. The second chart, spanning the same period, showed that the percentage of industry R&D funding allocated to long-term university research has also remained about the same.[31]

---

[29] OECD, Main Science and Technology Indicators, 2009.
[30] National Science Board, Science and Engineering Indicators 2010, Arlington, VA: National Science Foundation.
[31] Ibid., Science and Engineering Indicators 2010.

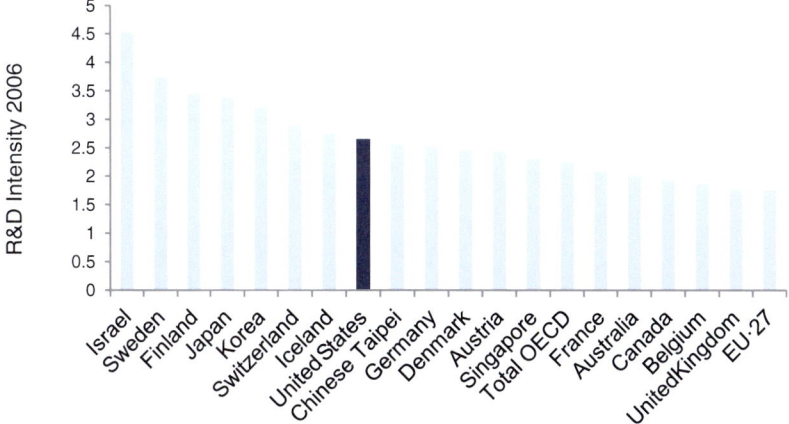

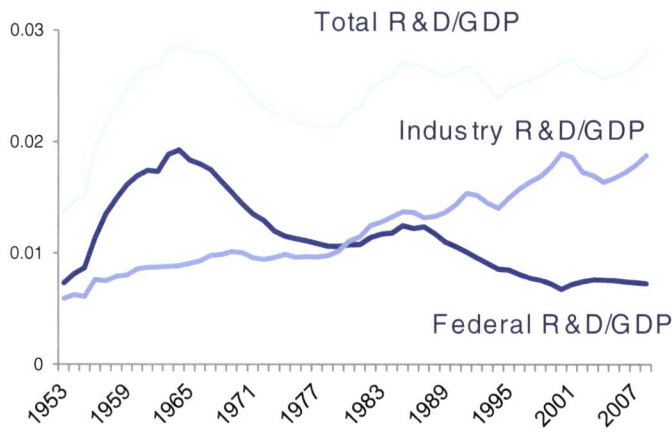

FIGURE 2 Problem: There are disturbing trends in R&D investment
SOURCE: Marc Stanley, Presentation at March 8-9, 2010 National Academies Symposium on "Building the Arkansas Innovation Economy."

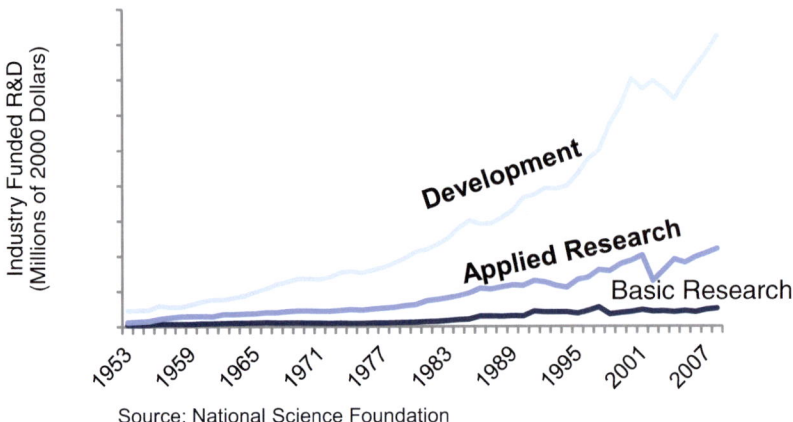

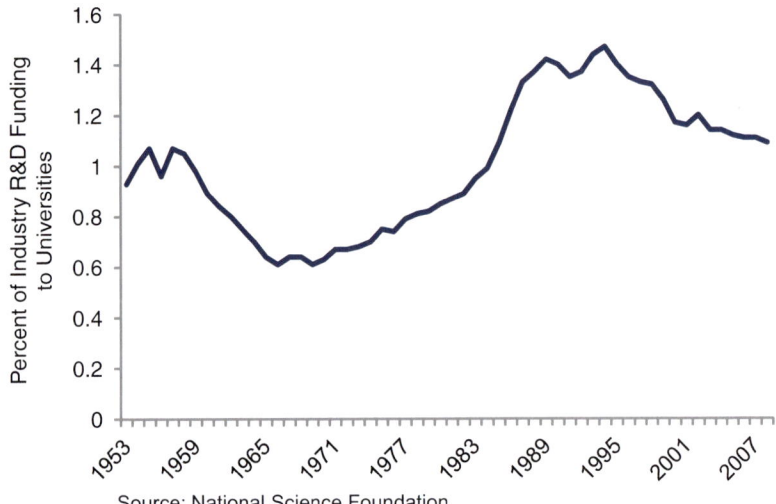

FIGURE 3 Problem: There are disturbing trends in R&D investment…and R&D Composition is Changing.
SOURCE: Marc Stanley, Presentation at March 8-9, 2010 National Academies Symposium on "Building the Arkansas Innovation Economy."

He said that despite much discussion at the federal level about increasing support for R&D, virtually no legislative changes have moved to the appropriations stage. On a policy level, he said, the most significant need was for federal agencies to move beyond their restricted silos of activity to more collaboration with other agencies with similar objectives. He noted some movement in this direction during the present administration, and urged much more.

Key Areas for Collaboration

The key areas for collaboration, he said, were regional policy, economic and industry policy, education policy, and science and technology policy. For example, he said that NIST, through the MEP, would have a vital but limited role in the energy regional innovation cluster. "We have to see if we can get out of our silos and help states like Arkansas keep their students here, grow companies, raise good revenue, and help our country grow." He said he was impressed with Arkansas' efforts to support its own innovation activities, especially in the face of a severe economic downturn.

He also said that he was proud of the administration's plans for the new RICs, and said that NIST had been building its own "RIC" model since 2007. The NIST version is call the Rapid Innovation and Competitiveness (RIC) Initiative, and its objectives are:
- Increase the nation's return on its R&D investments,
- Collapse the time scale of technological innovation,
- Encourage investment in need-driven research, and
- Stimulate the economy and enhance competitiveness.

This NIST RIC would be a public-private partnership for R&D investment, and began with reaching out to industries that already had a roadmap and helping them determine the research needs that would allow them to follow that roadmap. "This is highly industry-led," he said, "and I would highly recommend it for anything you're going to do in Arkansas."

The second objective of the RIC is research and knowledge transfer. Here, industry and government jointly fund goal-oriented basic research, measurement techniques, and the development of standards based on the needs and priorities of the roadmap. "As the governor said," he noted, "you have to get the universities involved, because that's where the gold is." Tacit knowledge transfer would be further facilitated by postdoctoral fellowships and personnel exchanges.

The third objective was to expedite the transition of scientific findings into commercial products. This may involve several approaches, such as helping companies find support as they cross the valley of death, and seeking to create a framework of support from state governments, regional organizations, and venture capital firms. It also paid attention to the need to evaluate each effort, which he called "a real failure on the federal side as well as the state side."

Launching the NIST RIC initiative

When NIST launched its RIC initiative in 2007, it began with a pilot Nanoelectronics Research Initiative (NRI), which is part of the Semiconductor Research Corporation (SRC). As NIST examined what its role should be, along with industry, states, and universities, it decided to design a collaborative model that could embrace organizations across the country. A five-year cooperative agreement was signed, with NIST bringing primarily its expertise in measurements and standards. Four regional centers were selected, involving 35 universities from 20 states, through a partnership with the NSF. The contributions of the states included tax incentives, research parks, and grants to nurture SMMs through the commercialization process. The program also support 128 graduate students and 24 post doctoral scholars at the four regional centers. The partnerships invested about $25 million per year: $2.75 million from NIST, $15 million from the states, and $5 million from industry, including venture capital and direct investment.

For this investment, the partnership realizes more than $200 million per year in business start-ups, development, and commercialization. During the first two years it has generated 13 patent applications and 239 scientific publications.

Mr. Stanley read two quotes from supporters of the program:

> "There is tremendous interest in every part of the world to win the nanoelectronics race and reap the economic rewards that will go with it. For America to win, it will take radical collaboration between government, higher education, and industry. The best example of this type of collaboration is the important work going on in the Nanoelectronics Research Initiative at more than 30 universities with funding and participation from NIST, IBM, and other major semiconductor and research institutions."
> - *John E. Kelly III, IBM Senior Vice President and Director of Research*

> "The NRI experiment is working; we learned more about graphene for device applications in the last two years than we would normally learn in 5 or 10 years in the business-as-usual research model."
> - *Industry member at INDEX review (9/08)*

In conclusion, he said, two current initiatives at NIST are (1) to press toward an even more "lean, agile" organizational structure, and (2) to ensure transparency in how taxpayer dollars are spent. The objectives of NIST will continue to be needs-driven basic research, linkage of NIST activities with those at the universities and regions, and continuing the Institute's world-class R&D on measurements and standards. He closed by encouraging Arkansas to look for opportunities to join a RIC, given the success of the program and the benefits to regional participants.

UNIVERSITY-FEDERAL GOVERNMENT PARTNERSHIPS

Donald Senich
Division of Industrial Innovation and Partnerships
Directorate of Engineering
National Science Foundation

Dr. Senich, senior advisor for small business procurement and research at the NSF, said that he would discuss NSF resources for the innovation ecosystem. A key to this approach, he said, was the effort of NSF director Arden Bement to better integrate research and education within the foundation. "That's the only way to get innovation," said Dr. Senich. "And that's why I think we're seeing a big change at the NSF."

He said that this year the Administration and Office of Management and Budget had put money into the NSF budget for innovation partnerships, intended to strengthen the innovation ecosystem. The important elements of that ecosystem, he said, included the universities, which were the source of new knowledge, but it also included other elements that were essential in promoting translational research. This approach, he said, had grown in importance across the NSF.

Funding Innovation is Expensive

Funding an innovation ecosystem, he said, was expensive, and the Recovery Act brought opportunities to accelerate critical ecosystem functions. But planners had to move both quickly and reasonably to allocate those resources. "Where could we quickly make a difference?" he asked. "With industry and the states. But how could we partner with them? We knew we had to grow our resources through translational research and working with partners, but we had to understand how that worked. That meant the right science and innovation policy.

"You can have all the research and education you want, but until you establish a well-grounded idea of innovation policy you're not going to go any place," said Dr. Senich. "The goals of this innovation ecosystem were to grow the existing portfolio and strengthen the translational phase; extend the reach of industry-driven initiatives; and better understand the social dimensions of innovation."

He acknowledged that innovation has many definitions and multiple elements. For this discussion, he defined innovation as a "new process, product, or service directed toward a social or economic change." He said that for the NSF, innovation could be translated as research that leads to quantifiable economic benefits. And part of the effort was to build sound metrics into an innovation program so those benefits could be measured.

Innovation at NSF

While most of the NSF budget is directed toward university-based research of a fundamental rather than results-driven nature, he said that NSF had, over the course of recent years, established a range of programs that involve university-industry partnerships, are outcome-driven, and contribute substantially to innovation. Successful examples, he said, included the Engineering Research Centers (ERCs), Industry/University Cooperative Research Centers (I/UCRCs), Partnerships for Innovation, and the Small Business Innovation Research (SBIR) and Small Business Technology Transfer (STTR) programs.

Describing an innovation spectrum from discovery (on the left) through the "Valley of Death" to commercialization (on the right), he noted that most of NSF's traditional activities in scientific discovery lie on the far left of the "Valley of Death." With regard to support for translational research, the major NSF programs were distributed across the spectrum from discovery to development according to their primary emphasis. He placed the Science and Technology Centers (STCs) and Materials Research and Science and Engineering Centers (MRSECs), with "the fundamental research activities" on the left. In turn, he placed the ERCs, then the I/UCRCs, and finally the SBIR programs successively on the right side of this spectrum, towards development and commercialization of new technologies.

He said he wanted to talk in more detail about "the space between the ERCs and the I/UCRCs." There were about 23 ERCs, he said, with 50 partners. They pursued six- to 10-year programs funded at about $3 million a year. There were 53 I/UCRCs, two of them in Arkansas, with a total of 135 partners. There were about 31 Materials Research Science and Engineering Centers (MRSECs), which "do not have partners but instead become groups." The most fundamental work was done by the STCs, of which there were 17, with 63 university partners.

'Clear Economic and Social Benefit'

The translational research within this mix, he said, is interdisciplinary by nature, involves teams, and relies on institutional partnerships. The distinctive quality of translational research, he said, which may or may not be present in basic, applied, or even developmental research, is "clear economic and social benefit."

Arkansas had received its own translational grants from the NSF, including some 20 awards in the SBIR (12) and STTR (8) programs between 2007 and 2009. These programs had brought a total of $4.25 million to the state.

Fifteen of the awards were made to the University of Arkansas in Fayetteville, jointly with its center of excellence in education. The key features of NSF centers include:
- A culture that joins research, education, and innovation,
- Ability to develop a diverse, globally competitive workforce,
- Production of creative and innovative practitioners to lead teams,
- Ability to leverage NSF funds to support industry-relevant research, and
- Partnership with industry to speed implementation and technology transfer

Most centers, he said, have as many or more partners as they have lead institutions.

In Arkansas, NSF supports an I/UCRC for engineering logistics and distribution that is a joint activity between the University of Arkansas and Sam's Club. While it is based in Arkansas, it has 11 partner groups at universities around the country. The center has created an Excel-based simulator to replicate the functionality of the Sam's Club inventory and logistics software, and has so far achieved a more than 4 percent reduction in inventory costs. Savings are expected to be as much as $70 million annually.

Overall, the NSF has classified its centers according to six technology categories: advanced electronics, advanced manufacturing, advanced materials, biotechnology, energy, and information technology. All are located at leading academic institutions. And over all, he said with enthusiasm, they deserve to share the same goal: "Educate to innovate." He closed by saying, "If I seem excited about what I'm doing, I am."

FROM UNIVERSITY RESEARCH TO START-UPS: BUILDING DEALS FOR ARKANSAS

Michael Douglas
UAMS BioVentures
University of Arkansas for Medical Sciences

Dr. Douglas said that the objective of his organization was "building deals for Arkansas," and that he would offer a picture of UAMS BioVentures by touching on "the numbers, the process, best practices, state incubators, and results."

He defined UAMS BioVentures as a "biomedical and biotechnology incubator" for the University of Arkansas for Medical Sciences. UAMS itself, located in Little Rock, is the home of the state's only teaching and research hospital and hub of advanced basic and clinical medical research in Arkansas.[32]

UAMS BioVentures focuses on inventions by UAMS researchers and ways to maximize the impact of those inventions. He said the financial impact of university inventions in the United States from 1996-2007 had reached $187 billion on U.S. gross domestic product, and a total of $457 billion on the U.S. gross domestic output, generating some 270,000 jobs. Based on surveys by the Association of University Technology Managers about one-third of invention disclosures from federal funding to universities had become patents. Of the total impact, about one invention disclosure was generated per $2.3 million of funding to universities, and one university start-up was generated per $50 million of federal grant support. "We as a country generate two tech companies a day just from National Institutes of Health funding," said Dr. Douglas.

He attributed much of the impact of university inventions to the Bayh-Dole Act of 1980,[33] which allows universities, such as UAMS, to retain ownership of their federally funded research and seek private investment to commercialize their discovery for better health care. The Act has been described by key government and business leadership as "one of the most important pieces of legislation in the 20th century."

The Complex Innovation Cycle

The process of commercializing even the most promising ideas is complex. Dr. Douglas described it as the innovation cycle, which he said created a "churn" of activities – knowledge creation (conception), technology transfer (formation), clusters and networks (maturity), and commercialization (growth). Because this churning process is fraught with resource limitations and uncertainties, he said, there is a need nationwide for incubators that can connect technology, people, and capital – to "fill the gaps" in commercialization. The Arkansas incubators, he said, perform that function through a number of mechanisms, including:
- Entrepreneurial training for the life scientist,
- Programs to train students to write and present business plans,
- Start-up advisory resources for early stage companies,
- A private equity roundtable network that extends over a seven-state area,

---

[32] http://www.uamshealth.com/.
[33] Also known as the University and Small Business Patent Procedures Act, Bayh-Dole was notable in reversing the presumption of intellectual property control from the federal government to individuals, small firms, or non-profits wishing to commercialize the results of their own federally funded research.

- Innovate Arkansas, a state-wide technology commercialization resource,
- Arkansas Development Finance Authority with its risk capital matching fund and tax credit programs, and the
- Research Park Authority, a new state authority to develop and support the acceleration of technology based companies and work force development.

These overlapping resources, he said, are viewed as elements in a "start-up continuum" extending from the university research investigator to the entrepreneur and venture funding within available capital and resource networks. "We have a small but well informed overlap," he said, "and cooperation among the various networks within the State."

A Profile of UAMS BioVentures

He then gave a profile of UAMS BioVentures as of 2009. The firm had two primary missions: (1) patent and licensing UAMS research and (2) starting technology-based biomedical companies.

Under the first mission, the company reported 296 patents or pending patent applications, 206 patent applications licensed, and 59 license and faculty support agreements. Royalty income from patents was about $1 million a year.

Under the second mission, the firm had helped form 12 companies that were currently operating, plus 22 client and pipeline companies. It had also set up Business Plan Teams at three state university campuses. In 2008, the companies paid a total of $21.5 million in salaries to some 420 employees.

In 2009, there were five business incubators in Arkansas, the first formed in 1986. (By comparison, he said, there were only two in the larger St. Louis metropolitan area with about double the population.) The incubators offered a total of 54,000 square feet of space to some 70 client companies and 33 operating companies. In Arkansas the incubators tend to be near the research universities in the northwest, northeast and central parts of the state.

He said that the economic productivity of the incubators was high, with an average annual wage of $56,000. The total capital raised by the incubators in Arkansas was $247 million, and the number of jobs created was 1252. He summarized the economic impact of UAMS BioVentures by the Institute of Economic Advancement, as of 2009, in the following terms:

- 44 start-up projects, with 22 in the portfolio and 19 in the impact study,
- $6.4 million in state and federal taxes generated,
- $76.3 million in total funding (debt, equity, and grants) to BioVentures' companies since 1997,
- $29.4 million in revenues (2008) from new products, services, and research,

- $52.4 million in 2008 economic output impact of BioVentures' companies, with 13 percent of the total out of state, and
- $184 million in total economic impact (1997-2008) as sales, investment, and research in the state.

He closed by characterizing with optimism the road ahead for commercialization of university research. The state has remarkable strengths in its strong business, government, and social networks, he said, and a "can-do approach" that was found in all of these sectors. Some challenges were to do a better job in "branding the region," building proof-of-concept funding resources, and expanding entrepreneurial networks. In particular, he noted, the state was "lacking" in adequate space for commercialization of research and technology parks to accelerate and recruit additional further technology development.

Session IV:
Universities and Regional Growth

Moderator:
John Ahlen
Arkansas Science and Technology Authority

Dr. Ahlen welcomed the final session, noting that every initiative discussed during the program, whether in Washington or Arkansas or other states, was created after 1980. "We only have three decades of experience doing this," he said. And while it had been "a slow, evolutionary process," it has "changed the path of economic development in Arkansas."

At the heart of that process, he said, was "clustering" on many levels, including the location of agencies. He said that his office had moved into a different building two weeks earlier, and for the first time in 30 years, all three economic development agencies were located in the same building. He recalled that one afternoon the directors of the other two walked into his office and began an impromptu 15-minute meeting "on something fairly important" that would not have happened without physical proximity. He suggested that technology clustering, public-private partnerships, and the many variations of collaborative work were equally important to S&T-based economic development.

ARKANSAS STEM COALITION ACTIVITIES

Michael A. Gealt
University of Arkansas at Little Rock

Dr. Gealt, dean of the College of Sciences and Mathematics and president of the Arkansas STEM (science, technology, engineering, and mathematics) Coalition, said that the coalition exists to promote excellence in STEM teaching and learning through a variety of strategies. It is a statewide partnership, recruiting members from the corporate, public, educational, and community sectors to ensure broad representation. And while its base of activity is educational, its broader purpose is to expand the economy of Arkansas and develop more high-paying jobs.

He described the urgency of his mission in terms of the "risk of inaction." The per capita income in Arkansas has been "too low" for too long, he said. Although it had risen from a low of about 36 percent of the national average in 1930 to approximately 75 percent of the national average in 1970, it had remained at that level for four decades.[34] He blamed this at least in part on poor performance in STEM subjects, both as taught and as learned.

"We can either stay where we are, or go down some more – or we can increase our capability in the STEM area and move up," he said. The goal of his organization, and of the Arkansas Research Alliance and Accelerate Arkansas, was to raise the per capita income to 100 percent of the national average.

Increasing Capability in STEM

He showed a graph illustrating a close relationship between a person's level of education and annual earnings.[35] "The more education you get, the more you'll be a lifelong learner." As state programs attempt to accelerate the economy, he said, it will be necessary to strengthen worker expertise in several ways. Higher education faculty and students must create more intellectual property that can lead to new industries. And developing a critical mass of experts in the higher education system will require that the pipeline of qualified students be filled by students in the pre-K-12 schools.

He said that the STEM coalition was concerned with every phase of education, from pre-K-12 learning through higher education and research. "We want to get them into the university system so they become the entrepreneurs of the future," he said. Of the Arkansas population, only about 20 percent are college-educated, compared with 26 percent of the national population. Current workforce needs included more emphasis on STEM, he said, because the knowledge-based economy of the future would require substantial STEM knowledge. Even workers who do not move beyond high school will require sufficient science and math skills to find employment in the general workforce.

Accordingly, the coalition seeks to promote education skills that strengthen the entire workforce. It does this by supporting high-quality STEM education, including studies in technology. The coalition also serves as "a think-tank for ideas about how to improve STEM education, and an advocate for educational models that have proven successful. It lobbies to influence educational policy, public understanding, and public engagement in STEM education."

"These successful models need both exposure and support if we are to adopt methodologies that have been shown to be working elsewhere. While it may be necessary to adopt them to Arkansas specifically, it is easier to build on the methods of others than to create a totally new process."

---

[34] Source: U.S. Bureau of Economic Analysis.
[35] Department of Labor data published in the 2004 Annual Report of the Federal Reserve System.

Integrating STEM Elements into the Core Curriculum

Among the specific goals of the Coalition are the integration of STEM concepts and technology elements into the core curriculum of grades P-16. "We have been major supporters of the state mathematics and science specialists who now work in state math and science centers and 'co-ops,'" he said. "Our efforts have supported the increase in the number of these math and science specialists and, in times of fiscal constraint, we have worked to maintain their number." The specialists have the roles of increasing enrollment in STEM courses in secondary education and educating parents, students, counselors and educators on the importance of STEM education. Additional goals are to increase enrollment in STEM courses in secondary education and develop better metrics of success.

Among the Coalition's accomplishments to date were funding packages for elementary science specialists (the state now has 27) and a web portal for STEM-related lesson plans. It has also supported legislation for grants to STEM teachers to increase their salaries through private funding. In 2007, the Coalition and Gov. Mike Beebe held a one-day conference to address educational issues and the need to build and sustain a workface with greater STEM skills. "I've heard the governor speak at length about the importance of education," he said. "This is a rare message to hear from a governor."

An important current activity, he said, is an effort to secure differential pay for STEM educators. He described a plan to formalize the coalition through 501(c)3 status and secure permanent funding.[36] Other objectives included a series of initiatives for the 2011 legislative session, elimination of the "opt out" option for the Smart Core Curriculum, securing differential pay for STEM educators, creating an elementary Science STEM Education degree program, and enhancing the technology infrastructure in Arkansas schools, including the educational portal for STEM educators.

At the core of the STEM coalition activities, he concluded, is an effort to better understand how to teach students and how to train teachers so their graduates have stronger STEM skills. The coalition is pursuing many avenues to

---

[36] According to the Internal Revenue Service, "To be tax-exempt under section 501(c)(3) of the Internal Revenue Code, an organization must be organized and operated exclusively for exempt purposes set forth in section 501(c)(3), and none of its earnings may inure to any private shareholder or individual. In addition, it may not be an action organization, i.e., it may not attempt to influence legislation as a substantial part of its activities and it may not participate in any campaign activity for or against political candidates." Access at http://www.irs.gov/charities/charitable/article/0,,id=96099,00.html/

improve these activities, including a close study of what is happening in other states and discussions with deans of education and P-12 educators.

STATE INITIATIVES FOR RESEARCH FUNDING AND THEIR ROLE IN ECONOMIC DEVELOPMENT

William Harris
Science Foundation Arizona

Dr. Harris, president and CEO of Science Foundation Arizona (SFAz), began with praise for Arkansas' approach to innovation, saying that it is "almost the model of what we need in our states." He highlighted the quality of its executive leadership, commitment to excellence in K-12 STEM education, and the ability to motivate people to work together toward a common objective.

He said that he would speak about three topics, and the ways in which all of them related closely to the programs being carried out in Arkansas in harnessing the power of science and technology for economic development. First he would describe his recent experience in Ireland, and in Brussels, where he gained important perspectives on building S&T capacity. Second, he would summarize some of the challenges faced by the United States and the individual states in strengthening S&T at home. Third, he would describe his current activities on behalf of the state of Arizona.

Lessons Learned in Europe

He began by reviewing some of the lessons he had learned during five years in Europe, which he began with an assignment to strengthen science, technology, and innovation in Ireland. There he served as founding director general of Science Foundation Ireland (SFI) and was responsible for a five-year, $1 billion program of strategic R&D investments in a country of four million people.

To build SFI "*from the ground up,*" he said, he had borrowed freely from the work of Vannevar Bush's 1945 report, "Science, the Endless Frontier," from his own experiences at the National Science Foundation (NSF) and in his work with other federal agencies. At the NSF he had witnessed a detrimental "explosion" of paperwork and bureaucracy that resulted in people spending more time and money on the preparation and review of proposals than on the actual science itself. Therefore one of his goals in Ireland was to focus tightly on world-class excellence and performance while keeping bureaucratic activities to a minimum.

The Value of an Outside Perspective

"One of the lessons from our SFI experience translates well at the U.S. state level," he said. "Arkansas will benefit greatly if you have some talented outsiders on your university, research, and economic advisory boards. Outsiders will help to ensure that your focus stays on performance and on value for money. Otherwise, your groups may be constrained by history: by 'we always did it this way' or distributing resources for traditional political reasons rather than according to merit and performance."

The goal of SFI, he said, was to build a culture and infrastructure for competitive funding of world-class research in Irish universities and research institutions. Proposals were reviewed in the same manner as they were reviewed in the United States at NSF and NIH, except that SFI did not use reviewers who were based in Ireland. "We wanted demonstrably to avoid any appearance of conflicts of interest and we wanted to expose other leading researchers to the outstanding science and engineering in Ireland."

He added that changing how a bureaucracy functions, "is about culture change and focusing on what is really critical." In Ireland, he said the SFI Board, which always had 3-4 international members, helped maintain the focus on performance – "speed and brains" – by overtly keeping score of Ireland's progress. "Holding up this mirror worked. After several years the external score keeper was no longer needed. A self-sustaining momentum emerged." The values of progress and performance, in other words, were quickly internalized and then monitored from within the country. SFI also catalyzed productive connections between industry and academia as it focused on use-inspired research.

While working in Ireland, leaders of the European Commission invited Harris to chair the high level expert group that was established to define the new European Research Council ("Frontier Research: The European Challenge" High-Level Expert Group Report, February 2005). His experience with the European Commission in Brussels, however, was less satisfactory. There he found a more abstract approach, and a bureaucracy that was "relentless in pushing its ambitious agenda," a trait that he had already seen in the U.S. federal programs. "This needs to be revised," he said. "If we are not strategic in building the implementation tool with buy-in from our states, we may waste time and money and discredit the ideas that have potential to work for us economically." He contrasted the "monolithic, centralized R&D in federal programs" with the "speed and diversity" that states like Arkansas and Arizona can achieve with the benefit of sound leadership and strategy.

He summarized the most important lessons he learned in Ireland, which were to operate independently, demand world-class levels of research and education and avoiding dependence on a central government to carry the whole burden of R&D. He translated these into the following best practices:

- Invest strategically at the state level in university-industry partnerships,
- Operate with speed and flexibility, and work opportunistically,
- Strive for world-class standards for STEM K-12 performance/education,
- Build partnerships with industry, and
- Listen to R&D-driven business entities and become IP-friendly.

Speed and Flexibility

As he reviewed his European experiences in the light of U.S. S&T strategies, he saw a series of challenges. First, he said, the times had changed radically since the Cold War era when Vannevar Bush helped create the U.S. science funding structure. Today, the U.S. faces dozens of new economic competitors around the globe and significantly stronger universities in Asia and Europe. "Our bureaucracies need refinement" he said, "if not wholesale redesign at the state and federal levels." In Ireland, "the emphasis was on getting things done," and government agencies acted to ensure that industry could be fully successful. "In the U.S., unfortunately beyond national security, we seem less able to focus and recognize the importance of working together to advance our economy and education systems. We are *drifting* and we may be missing the opportunity to lead critical sectors, such as energy technology."

Links with Business

Our bureaucratic systems tend to lose contact with our industries and their challenges, he stated. A historical model many people have forgotten, he said, was that of the public land-grant universities and their connections with the industry of the era. More recently, another kind of connection began with the Bayh-Dole Act of 1980, which gave universities and others incentives to exploit discoveries supported by government funding. After almost 30 years of experience with Bayh-Dole, however, more agility is required to transform innovations into valuable public uses and private advantage. In Ireland, R&D leaders from U.S. industry were eager "to work with Irish universities, because they and the SFI seemed less bureaucratic than U.S. counterparts – and because Irish legislation supported the exploitation of discoveries."

"In an ideal world," he said, "the states would take the initiative to invest in strategic R&D in response to strategic priorities – just as Ireland did. However, in the current economic crisis, a Federal investment program may be needed to stimulate strategic state R&D initiatives."[37]

---

[37] William C. Harris, "Innovation lessons from Ireland," *Research-Technology Management*, Volume 53, Number 1, January-February 2010. pp. 35-39(5).

The Arizona Challenge

After Dr. Harris returned from his time in Europe, he was invited in 2006 to apply what he had learned in Ireland to Arizona, at the invitation of Arizona Governor Janet Napolitano and significant business leaders (i.e., Greater Phoenix Leadership). "Research is mobile," he said. "Arizona is an example of a new model of open innovation. The governor and the business communities were interested in trying to learn how the government can connect universities to business in new ways. We are finding that the best way to do this is to work in partnership with individual states."

Governor Napolitano, he said, was concerned that although the state had grown rapidly population wise, the economy was still not diverse enough to sustain the growth over time. Arizona had many new people and many retired people, some of them working part-time. The challenge was to update and diversify the economy and to enhance the focus on education standards. "There were similarities with Ireland," he said, "and Arizona's leaders decided that Science Foundation Ireland was the right model for Arizona."

The stated purpose of Science Foundation Arizona (SFAz) was "to diversify and strengthen Arizona's economy," sufficiently that it could compete in the global marketplace. It would do this through a four-pronged strategy:

A Public-Private Partnership

The new organization – *Science Foundation Arizona* – was structured as a public-private partnership, with the business community, including philanthropic leaders, and the state whom each provided half of the funding. Almost all of the funding – $100.9 million[38] – was designated for awards, with only $1.78 million annually allocated to operating costs that was funded by the 3 CEO Groups.[39] The awards were distributed based on external peer review – not regional or political input – and by the type of activity, with just over half (56 percent) going to "strategic research." Of the remainder, 17 percent went to graduate research fellowships, 9 percent to competitive advantage awards, 5 percent to catalyze small business growth, and 12 percent to K-12 "teacher and student discovery."

Awards were focused on areas judged to be of high priority to Arizona's citizens. These included solar and wind energy, sustainable mining,

---

[38] Committed and actual funds as of January 2010.
[39] Greater Phoenix Leadership (GPL), Southern Arizona Leadership Council (SALC) and Flagstaff 40 (Flag40).

personalized medicine, and new materials and software related to computer chip and aerospace sectors. Some specific examples of new R&D partnerships were:

- The Institute for Mineral Resources at the University of Arizona, which has 15 significant industry partners
- The Solar Technology Institute, which has initiated programs ranging from concentrated solar energy to energy storage
- A joint investment with Arizona State University to transform algae into jet fuel and a spin-off company to demonstrate the value of the fuel for private airlines and the Department of Defense;
- The Critical Path Institute (C-Path) to transform drug development through the work of its three consortia, which include more than 500 scientists from 30 global pharmaceutical companies, the FDA, and its European counterpart, EMEA.

In June 2009, after about two years of SFAz activities, Battelle evaluated its return on investment, and found $2.18 in new monies were leveraged from each $1 awarded in university grants. It also found other outcomes: a "STEM education impact" on 54,000 students and 680 teaches, 11 spin-off companies, 757 jobs created or retained, 50 patents filed or issued, and 292 scientific publications.

STEM K-12 Education

"While the federal government has the prime responsibility for our research infrastructure, the states and localities have prime responsibility for the K-12 system. Although the K-12 education system is a key to our competitiveness and well-being, U.S. high school graduates do not rate highly in science and mathematics compared to our global trading partners. Many Asian countries are producing extraordinarily able scientists and engineers; we can no longer count on their top talent moving here."

An expressed state R&D investment strategy will inform state legislators about the need to focus on STEM in K-12. SFAz has invested in connecting education to hands-on experience so students understand the "*what and why*" that is behind course work. The state focuses on growing its own talent pool, and supports a program with area businesses to improve science and math teaching and gives teachers real-world experience" in the summer. "Our pilot programs are successful," he said, "because of the concern the business community has for the region's future."

Dr. Harris closed with a proposal for Arkansas and for the federal government. "Though many things have changed," he said, "we continue to believe that R&D is an 'endless frontier' for the United States. An alternative world, in which the frontier is closing, is unacceptable. Yet to expand our horizons and to gain ground will require bold experiments at every level. We have the potential to retain our strength in education, research, and innovation –

but we need to suit up and compete in the 21st Century global system and not accept mediocrity. The nation would be well served by a Federally initiated series of competitive pilot programs – perhaps in 10 to 15 states – to encourage innovation by linking the business community with the universities and other strategic assets of the states in new ways. Arkansas seems well prepared to do just that."

Session V:
Arkansas R&D Capacity:
Universities Research Labs and Science Parks

Infrastructure for High-Performance Computing

Amy Apon
High-Performance Computer Center
University of Arkansas, Fayetteville
Division of Computer Science, Clemson University

Dr. Apon, then a professor of computer science and Founding Director of the Arkansas High-Performance Computing Center, said that her work had been strongly supported by the state of Arkansas and by the NSF.[40] The focus, she said, was cyberinfrastructure, which she said could be defined as the "IT infrastructure that enables scientific enquiry." A statewide task force had developed a plan for cyberinfrastructure, with three entities:

The first was ARE-ON, the Arkansas Research and Educational Optical Network, an initiative to connect all the four-year public institutions in the state to a 10Gb network and provide access to state resources for anyone at any four-year institution. ARE-ON was scheduled to be fully operational at the end of the current semester, providing full access to national cyberinfrastructure resources. "The key message," she said, "is that we are part of a national ecosystem." She showed a list of collaborators in Texas, Louisiana, Pennsylvania, and elsewhere. "We can all share resources from our desktops at our four-year institutions."

A Correlation between Federal Funding and Computational Capacity

Second, the Star of Arkansas was the state's largest computational resource, she said, funded through an NSF grant that provided about 11 million computing hours per year. "There is a lot of research in Arkansas that can benefit from computation," she said. "One area is complex data analysis using emerging technologies to analyze data much more rapidly." She added that there is an 80 percent correlation between a state's level of federal funding for computation and the state's ranking in computational capacity.

Another field in which computation is central is the accurate description of large molecules. She mentioned the work of Peter Pulay, a

---

[40] Dr. Apon is currently affiliated with the School of Computing at Clemson University.

distinguished professor of chemistry at the University of Arkansas, Fayetteville, who studies the interaction of chemicals on human protein and DNA structure. Pulay's research requires 4 million hours of computing time each year.

Another computing challenge is found in nanotechnology, where scientists have already reached fundamental limits in computer technology. Research by Laurent Ballaiche has the potential to create nanotechnology devices that can build memory 10,000 times denser than anything currently manufactured, and his research requires 70 million hours of compute time per year. This is more than can be supported in Arkansas, she said, which is "why it's important to have access to national resources."

In materials science, the challenge is to model plasticity and failure in metal alloys, with applications in aeronautics and other fields. Doug Spearot, assistant professor of mechanical engineering, creates 3-dimensional models of alloys that do not yet exist, using 20 million or more atoms. The computers evaluate variations of the alloys before they are fabricated in a laboratory. This modeling study requires 6 million hours of compute time a year.

The Most Important Instrument of Science

She emphasized the importance of supercomputing with a quote from Jay Boisseau, director of the Texas Advanced Computing Center: "Over the past 60 years, computing has become the most important general-purpose instrument of science." Dr. Apon added, "My dad used to tell me, mathematics is the language of science. Well, computing is the most important general purpose instrument of science."

In addition to ARE-ON, she said, NSF now funds a new EPSCoR Track 2 project called CI Train, or Cyberinfrastructure for Transformational Scientific Discovery. The intention is to provide campus cyberinfrastructure "champions" to serve as liaisons between the physical resources and the researchers and educators who need access to them. It also provides visualization resources and supports research in a wide spectrum of computational and visualization domains.

In Arkansas, she said, the CI Train project has made education a "key deliverable," with partners across the state. It supports initiatives at the high school, undergraduate, and graduate levels with professional information technology staff and research faculty. It shared nationally competitive visualization resources and large-scale computational resources.

Re-thinking our Campus Environments

These new resources required unprecedented levels of sophistication for computer data visualization, she said. "We need to seriously rethink our campus environments and how they can support new data-driven modalities of research, collaboration, and education. She credited Rob Pennington of the NSF

Office of Cyberinfrastructure for bringing this perspective, and also demonstrating how technologies can help scholars and students talk to each other and leverage all the resources across the state.

She closed with several examples from the NSF that pair problems with solutions. For example, recent computer science PhD training is disconnected from what scientists need; similarly, recent physics PhDs are not trained in software engineering. An example of a solution: create post-doc-to-professoriate programs to encourage them to apply their knowledge (and protect them).

"There are many needs," she said in closing. "We must educate students at all levels in collaborative computational science. This is hard. Computer scientists don't necessarily want to do chemistry, and the chemists don't necessarily want to learn how to write programs on emerging technologies. It has to be a first-class, joint effort. We also have to encourage and support researchers moving into these areas, and we have to catalyze cultural changes in academics and agencies to better support interdisciplinary activities – not just by the faculty but also by the administration."

RESEARCH PARKS IN ARKANSAS

Jay Chesshir
Little Rock Chamber of Commerce

"We are trying," Mr. Chesshir began, "to take a state that has not necessarily been known for technology and innovation and move it into a brand-new world."

With that, he said, he wanted to talk about research parks in Arkansas and what the state is doing to grow them. There were currently two science parks: the Arkansas Research and Technology Park, adjacent to the University of Arkansas in Fayetteville, and the Arkansas Bioscience Innovation and Commercial Center at Arkansas State University in Jonesboro, which is completing its Phase I business incubator. A new park was being constructed in central Arkansas.

The Arkansas Research and Technology Park (ARTP) used innovative techniques to nurture technology-intensive companies. It attempted to stimulate the formation of a collaborative community of companies, together with university faculty and students at Fayetteville, linked interdependently around a set of core R&D research competencies at the university.

Growing our own Expertise

"We learned in the last several years," he said, "that people are not coming here in search of expertise. We're going to have to do a better job of growing it ourselves. We're not going to Boston, Ireland, or India to recruit that type of technology and innovation to come here and blossom. We are going to have to grow our own."

The ARTP is an example, he said, of a community that has begun to create the next generation of electronic and photonic devices for biotechnology and related areas. These areas include transportation and logistics, in which Arkansas is a leader; materials and manufacturing; database software; telecommunications; and applied sustainability. Those are areas in which the ARTP is successful in terms of grants attracted and progress toward becoming a center of excellence.

The State's Primary Knowledge Community

A major advantage for the ARTP, he said, was its location in northwest Arkansas, near the main university campus. As the state's primary knowledge community, the Fayetteville area provided valuable fuel for the innovation and technology development activities of the ARTP. Two affiliates had received the prestigious Frost and Sullivan Award for excellence in technology, and another affiliate won the Tibbetts Award for the most innovative small business. Earlier in the year, another affiliate won an R&D 100 award, which cites Washington County as one of the most innovative in the country. ARTP affiliates, he said, continue to advance the frontier of product development in many specialty areas.

The reason that is important, he said, is that "what's going on up there in northwest Arkansas permeates the state, and provides a sense of innovation for folks in the other universities."

In central Arkansas, a group had engaged a consultant to review activities at the University of Arkansas for Medical Sciences and the University of Arkansas at Little Rock. The question they asked was: How can we take the research and innovation that is already here and make it stronger? For so long, he said, the state had suffered from brain drain as its best and brightest young scientists, engineers, and medical researchers sought opportunities elsewhere. How, they asked, could the region take advantage of local innovative talent and turn it into jobs for the area and the state as a whole.

In 2007, this effort was rewarded when the General Assembly voted to create a research park authority, a legislative opportunity that would permit anyone in the state to create a research park and design it for sustainability. That effort had moved forward, he said, and at the end of the month, the authority was scheduled to be finalized with the city of Little Rock and its partners in central Arkansas, with the goal of beginning construction by 2012.

"This is a very ambitious schedule and investment," he concluded. "It is something that has never been done in central Arkansas. Only when we have all our state, academic, and government partners working together are we going to be as successful as we need to be."

UNDERSTANDING THE BATTELLE STUDY

Jerry Adams
Arkansas Research Alliance

Mr. Adams, who had founded the Arkansas Research Alliance after retiring from Acxiom Corporation, said that Battelle was asked to do a thorough study of economic development in Arkansas, primarily to provide evidence-based insight into the core competencies of the research universities. He said that at Acxiom, he had received many calls from people asking for funding for research that was unrelated to Acxiom's core activities. One lesson the company had learned, he said, was that it made sense to pay only for research that could move the country ahead. "So a key issue we discussed with Battelle," he said, "was what are we good at in Arkansas? What can move us ahead?"

Battelle did a qualitative review based on field interviews with 85 of the top researchers in the state, and a quantitative review based on the journal publications and research grants of faculty members during the last five years. "In other words, by looking in the rear-view mirror."[41]

Core Competencies and Economic Benefits

Nonetheless, he said, the study turned out to be a living document that revealed more than a dozen core competencies in Arkansas. But it also presented the challenge of finding the best way to derive more economic benefits from those 18 core competencies and 12 niche competencies.

"This was surprising to Battelle," he said, "but 30 turned out to be too big a number. So we rolled them up into nine strategic focus areas." Those were multi-disciplinary fields of research that were likely to enable the state to leapfrog more traditional universities that have more strength in narrow academic fields. The focus areas were also designed to engage multiple institutions, rather than be limited to individual universities or geography. The focus areas were:

- Enterprise systems computing,
- Distributed energy network systems,
- Optics and photonics,
- Nano-related materials and applications,
- Sustainable agriculture and bio-energy management,

---

[41] Battelle Technology Partnership Practice, "Opportunities for Advancing Job-Creating Research in Arkansas, A Strategic Assessment of Arkansas University and Government Lab Research Base," 2009. Access at http://www.aralliance.org/__data/assets/pdf_file/0017/1682/Job-Creating-Research-in-Arkansas.pdf.

- Food processing and safety,
- Personalized health research sciences,
- Behavioral research for chronic disease management, and
- Obesity and nutrition.

Increasing Multi-Campus Collaborations

"A reality of being a small southern state," he said, "is that we're about $106 million below where we should be on a per capita basis in attracting federal research dollars to the state. Part of that is due to a lack of multi-campus collaboration, Battelle learned, so that a theme adopted by the ARA was to raise the number and level of multi-campus collaborations around those core competencies."

The study looked at the competencies in terms of whether they were emerging, limited, or established, and examined the level of federal funding. "That is the accelerator," he said. "State funding can help support talent, but federal funding is the key." The next question was about market potential, the pull from industry, and whether there were already existing industries in Arkansas in these areas. He noted that Tom Dalton, of Innovate Arkansas, tried to "validate the technology: is this a business likely to stay in Arkansas, or will we create something that will move to Boston?"

He noted that the Arkansas Biosciences Institute already engaged in three of the core competencies, but that "we're not in competition with ABI, we're a partner with them." He said that nine areas would eventually be too many to focus on, but that "we needed an evidence-based roadmap like Battelle's, as opposed to hearing researchers tell us how terrific their research is. Self-reported results have bias."

A 'Crucial Roadmap' for Recruiting Talent

He said that the ARA had found the Battelle study to be a "crucial roadmap" to use in recruiting talent into the state, and into the core focus areas. For example, he said that Arkansas was in the process of launching an eminent scholars' program, modeled after one in Georgia.

The ARA was also trying to elevate the level of multi-campus collaboration, with funding from the Winthrop Rockefeller Foundation, the Walton Family Foundation, and the ARA board. With the help of administrative and research leaders of each of the five campuses, they had planned three conferences on (1) smart infrastructure, including the smart grid, (2) "smart information," and (3) nanotechnology. They were also planning a conference on "healthy Arkansas," and another based on bio-production and clean energy. The conferences would cover the nine focus areas during an 18-month period, with the goal of helping planners decide which to institutionalize in the state.

He closed by affirming that the "Battelle study has been a gift to this state. It's been absorbed into the EPSCoR conversation, and in the effort to focus the state's research resources. My hope is that we will use it as our investment roadmap going forward."

CLOSING REMARKS

John Ahlen
Arkansas Science and Technology Authority

Dr. Ahlen closed the symposium by exhorting his audience to go beyond the discussion stage and move into action. He observed that the Science and Technology Authority had a family of core programs and was managing two federal projects: an MEP project, with NIST, and an EPSCoR project, with NSF. The state and federal managers talked several times a week about these projects and other state activities. He asked, "How do we streamline these relationships?" NSF is primarily a research organization, he said, but it wants to see the results of research commercialized. "They tell the state we have to do that, and we do it through EPSCoR. But so do the state agencies that have been doing economic development for decades. The MEP would like to see more innovation in manufacturing. We applaud that, but we've also been trying to do that for 30 years.

"So it is time to look at these relationships, streamline them, and realize that we're all trying to move to the same place. We have multiple rules at the state and federal levels, and for those of us trying to execute, it's very difficult. All these rules are designed for transparency and accountability, but to different bosses in different places.

"I will remind some of us that 12 years ago, the National Science and Technology Council at the White House had its first interagency task force meeting on innovation partnerships, and after a couple of years the momentum came to a grinding halt. Here we are 12 years later having that same discussion. We don't have another decade to sit on this and wait for another discussion. We need to pick up the phone and call those friends in Washington who have told us to call.

"So," he concluded, "go forward and collaborate."

III

APPENDIXES

Appendix A

Agenda

Building the Arkansas Innovation Economy

A Symposium Organized by
The U.S. National Academy of Sciences
in cooperation with
The University of Arkansas at Little Rock

March 8-9, 2010

Clinton Presidential Center
Little Rock, Arkansas

DAY 1

2:00 PM **Welcome and Introductions**
Mary Good, University of Arkansas at Little Rock

2:15 PM **Session I: The Global Challenge and the Opportunity for Arkansas**
Moderator: Mary Good, University of Arkansas at Little Rock

The Innovation Imperative: Global Best Practices
Charles Wessner, Director, Technology, Innovation and Entrepreneurship, The National Academies

Innovation Infrastructure at the State and Regional Level: Some Success Stories
Richard Bendis, President and CEO, Innovation America

Innovation and Commercialization Successes in Oklahoma
David Thomison, Vice President of Enterprise Services, Innovation to Enterprise (i2E)

California's Innovation Challenges and Opportunities
Susan Hackwood, Executive Director, California Council on Science and Technology

Evolution of Innovation in Arkansas
Watt Gregory, Executive Committee Chair, Accelerate Arkansas

4:00 PM **Session II: Cluster Opportunities for Arkansas**
Moderator: Paul Suskie, Chairman of Public Service Commission

Arkansas and the New Energy Economy
Paul Suskie, Chairman of Public Service Commission

Federal-State Synergies
Gilbert Sperling, Energy Efficiency and Renewable Energy (EERE), U.S. Department of Energy

The Wind Energy Industry in Arkansas: An Innovation Ecosystem
Joe Brenner, Vice-President for Production, Nordex USA

Arkansas's Role in Energy Transmission Management
Nick Brown, President and CEO, Southwest Power Pool

5:40 PM **Adjourn**

DAY 2

8:30 AM **The State of Technology and Innovation in Arkansas**
The Honorable Mike Beebe, Governor of Arkansas

8:50 AM **Session II: Cluster Opportunities for Arkansas (continued)**
Moderator: Charles Wessner, Director, Technology, Innovation and Entrepreneurship, The National Academies

Research in Advanced Power Electronics: Status and Vision
Alan Mantooth, Director, University of Arkansas's National Center for Reliable Electric Power Transmission (NCREPT), University of Arkansas at Fayetteville

APPENDIX A   *145*

                       Regional Innovation Clusters (RIC)
                       Ginger Lew, National Economic Council, The White House

                       Agriculture and Food Processing
                       Carole Cramer, Director, Arkansas Biosciences Institute, Arkansas State University

                       Information Technology
                       Jeff Johnson, President and CEO, ClearPointe

                       Nanotechnology
                       Greg Salamo and Alex Biris, University of Arkansas at Fayetteville

10:10 AM   **Session III: Federal and State Programs and Synergies**
                       Moderator: Barry Johnson, Senior Advisor and Director of Strategic Initiatives, Economic Development Administration, Department of Commerce

                       The Role of the Economic Development Administration
                       Barry Johnson, Senior Advisor and Director of Strategic Initiatives, Economic Development Administration, Department of Commerce

                       Initiatives of the Manufacturing Extension Program
                       Roger Kilmer, Director, Manufacturing Extension Partnership, National Institute of Standards and Technology

                       University-Industry Partnerships
                       Marc Stanley, Deputy Director, National Institute of Standards and Technology

                       University-Federal Government Partnerships
                       Donald Senich, Division of Industrial Innovation and Partnerships, Directorate of Engineering, National Science Foundation

                       From University Research to Start-ups: Building Deals for Arkansas
                       Michael Douglas, Director, UAMS BioVentures, University of Arkansas Medical Services

11:30 AM   **Lunch**

12:15 PM	**Session IV: Universities and Regional Growth** *Moderator: John Ahlen, President, Arkansas Science and Technology Authority*
	Arkansas STEM Coalition Activities *Michael A. Gealt, UALR dean of College of Sciences and Mathematics, President of Arkansas STEM Coalition*
	State Initiatives for Research Funding and Their Role in Economic Development *William Harris, President and CEO, Science Foundation Arizona*
1:15 PM	**Session V: Arkansas R&D Capacity: Universities, Research Labs, and Science Parks** *Moderator: John Ahlen, President, Arkansas Science and Technology Authority*
	Infrastructure for High-Performance Computing *Amy Apon, High-Performance Computer Center, University of Arkansas at Fayetteville, and Division of Computer Science, Clemson University*
	Research Parks in Arkansas *Jay Chesshir, President and CEO, Little Rock Chamber of Commerce*
	Understanding the Batelle Study *Jerry Adams, President and CEO, Arkansas Research Alliance*
2:45 PM	**Adjourn**

Appendix B

Participants List*

Jerry Adams
Arkansas Research Alliance

Nitin Agarwal
University of Arkansas at Little Rock

John Ahlen
Arkansas Science and Technology Authority

Jenny Ahlen
AEDC

Alan Anderson
The National Academies

Gary Anderson
University of Arkansas at Little Rock

Joel Anderson
University of Arkansas at Little Rock

Amy Apon
University of Arkansas at Fayetteville

Jeff Amerine
Innovate Arkansas

Stan Baker
Stanley Baker, Ltd.

Ramsay Ball
Virtual Incubation Board of Directors

Governor Mike Beebe
State of Arkansas

David Belcher
University of Arkansas at Little Rock

Richard Bendis
Innovation America

Alex Biris
University of Arkansas at Little Rock

Joe Brenner
Nordex

Sen. Shane Broadway
Arkansas General Assembly

Nick Brown
Southwest Power Pool

Jay Chesshir
Little Rock Regional Chamber of Commerce

Charisse Childers
Accelerate Arkansas

---
*Speakers in italics

Tom Chilton
Arkansas Economic Development Commission

McAlister Clabaugh
The National Academies

Bill Clay
Acxiom Corporation

Carole Cramer
Arkansas State University

Tom Dalton
Innovate Arkansas

Ted Dickey
Innovate Arkansas

David Dierksheide
The National Academies

Michael Dockter
Arkansas State University

Michael Douglas
UAMS Bioventures

Gene Eagle
Arkansas Development Finance Authority

Tamika Edwards
Office of U.S. Sen. Blanche Lincoln

Sen. Joyce Elliott
Arkansas General Assembly

Laura Fine
Arkansas Small Business Development Center

Susan Forte

Ed Franklin
Arkansas Association of Two-Year Colleges

Michael Gealt
University of Arkansas at Little Rock

Collis Geren
University of Arkansas at Fayetteville

Adam Gertz
The National Academies

Patricia Gonzalez
U.S. Export Assistance Center Arkansas

Mary Good
University of Arkansas at Little Rock

Watt Gregory
Accelerate Arkansas

Susan Hackwood
California Council on Science and Technology

Maria Haley
Arkansas Economic Development Commission

William Harris
Science Foundation Arizona

James Hendren
Hendren Consulting

Vernard Henley
University of Arkansas at Little Rock

APPENDIX B

Brad Henry
Arkansas Economic Development Commission

Karin Iqbal
University of Arkansas at Little Rock

Jeff Johnson
ClearPointe Technology

Barry Johnson
U.S. Economic Development Administration

Roger Kilmer
National Institute of Standards and Technology

Karrie Kovalcheck
HP

Ginger Lew
National Economic Council

Mark Malone
Northwest Arkansas Council

Alan Mantooth
University of Arkansas at Fayetteville

Gary McChesney
Future Fuel

Gail McClure
Arkansas Science and Technology Authority

Dina Nash
University of Arkansas at Little Rock

Rebecca Norman
Arkansas Small Business Development Center

Leo Perreault

Janet Marie Roderick
Arkansas Small Business Development Center

Greg Salamo
University of Arkansas at Fayetteville

Mark Saviers
Sage Partners

Don Senich
National Science Foundation

Sujai Shivakumar
The National Academies

Mike Smith
Innovate Arkansas

Gilbert Sperling
U.S. Department of Energy

Phil Stafford
University of Arkansas at Fayetteville

Marc Stanley
National Institute of Standards and Technology

Mayor Mark Stodola
City of Little Rock

Paul Suskie
Public Service Commission

David Thomison
i2E

Becky Thompson
Arkansas Business Development Commission

Tom Walker
University of Arkansas at Little Rock

Sam Walls
Arkansas Capital Corporation

Charles Wessner
The National Academies

Deke Whitbeck
Office of U.S. Sen. Mark Pryor

Donna Kaye Yeargen
Office of U.S. Sen. Blanche Lincoln

Kenji Yoshigoe
University of Arkansas at Little Rock

Jim Youngquist
UALR Institute for Economic Advancement

Randy Zook
Arkansas State Chamber of Commerce

Appendix C

Bibliography

Accelerate Arkansas Strategic Planning Committee. 2007. *Building a Knowledge-Based Economy in Arkansas: Strategic Recommendations by Accelerate Arkansas*. Teresa A. McLendon, ed. Little Rock: Accelerate Arkansas.

Acs, Z., and D. Audretsch. 1990. *Innovation and Small Firms*. Cambridge, MA: The MIT Press.

Alic, J. A., L. M. Branscomb, H. Brooks, A. B. Carter, and G. L. Epstein. 1992. *Beyond Spin-off: Military and Commercial Technologies in a Changing World*. Boston: Harvard Business School Press.

American Wind Energy Association. 2011. "Arkansas is a National Leader in Wind Energy Manufacturing." Washington, DC: American Wind Energy Association. August.

Amsden, A. H. 2001. *The Rise of "the Rest": Challenges to the West from Late-industrializing Economies*. Oxford: Oxford University Press.

Arizona Daily Star. 2009. "Budget Cuts Hit Science Research Partnerships at Arizona Universities." February 8.

Arkansas Business. 1996. "Nucor Makes Blytheville Steel Capital of the South." December 16.

Arkansas Business. 2011. "Beebe, FDA Sign First of its Kind Agreement at NCTR." August 12.

Arkansas Business. 2011. "NCTR Has Potential to Create High-Paying Jobs." July 4.

Arkansas Business. 2012. "AMS Grant Helps Local Aerospace Manufacturer Turn Business Around." January 5.

Arkansas Democrat-Gazette. 1987. "Arkansas Legislators Present Their Proposal for Tax Breaks for Proposed Steel Mill." December 7.

Arkansas Democrat-Gazette. 1999. "Results from Subsidies Unknown–State Has Little Idea Whether $633 Million in Breaks to Firms Spurred Investment." December 12.

Arkansas Department of Education. 2007. *Combined Research Report of Business Leaders and College Professors on Preparedness of High School Graduates*. January. Little Rock: Arkansas Department of Education.

Arkansas Economic Development Commission. 1979. *Arkansas Climbs the Ladder: A View of Economic Factors Relating to Growth of Jobs and Purchasing Power*. Little Rock: Arkansas Economic Development Commission.

Arkansas Economic Development Commission. 2002. "Report of the Task Force for the Creation of Knowledge-Based Jobs." Little Rock: Arkansas Economic Development Commission.

Arkansas Economic Development Commission. 2009. *Governor Mike Beebe's Strategic Plan for Economic Development.* Little Rock: Arkansas Economic Development Commission.

Arkansas Research and Education Optical Network. 2008. "Arkansas Cyberinfrastructure Strategic Plan." Little Rock. <http://areon.net/resources/CyberinfrastructureStrategicPlan2008102 4.pdf>

Arkansas Small Business and Technology Development Center. 2009. "Enterprise Center to Offer Valuable Technology Incubator Resources." Press Release. Little Rock: University of Arkansas at Little Rock.

Arkansas State University. 2011. "Brian Rogers Named Director of Commercial Innovation Technology Incubator." Press Release. Jonesboro, AR: Arkansas State University. January 5.

Arkansas Task Force on Higher Education Remediation, Retention and Graduation Rates. 2008. *Access to Success: Increasing Arkansas' College Graduates Promotes Economic Development.* ("Education Task Force Report.") August.

Arkansas Task Force on Higher Education Remediation, Retention, and Graduation Rates. 2008. *A Plan for Increasing the Number of Arkansans with Bachelor's Degrees*. Little Rock: Arkansas State University.

ArkansasOnline. 2008. "LM Glasifiber Dedicates Little Rock Factory." October 28.

Asheim, B., A. Isaksen, C. Nauwelaers, and F. Todtling, eds. 2003. *Regional Innovation Policy for Small-Medium Enterprises*. Cheltenham, UK, and Northampton, MA: Edward Elgar.

Athreye, S. 2000. "Technology policy and innovation: The role of competition between firms." In P. Conceicao, S. Shariq, and M. Heitor, eds. *Science, Technology, and Innovation Policy: Opportunities and Challenges for the Knowledge Economy*. Westport, CT, and London: Quorum Books.

Atkinson, R., and S. Andes. 2010. *The 2010 State New Economy Index: Benchmarking Economic Transformation in the States.* Washington, DC: Kauffman Foundation and the Information Technology and Innovation Foundation. November.

Audretsch, D., ed. 1998. *Industrial Policy and Competitive Advantage, Volumes 1 and 2*. Cheltenham, UK: Edward Elgar.

Audretsch, D. 2006. *The Entrepreneurial Society*. Oxford: Oxford University Press.

Audretsch, D., B. Bozeman, K. L. Combs, M. Feldman, A. Link, D. Siegel, P. Stephan, G. Tassey, and C. Wessner. 2002. "The economics of science and technology." *Journal of Technology Transfer* 27:155-203.

Audretsch, D., H. Grimm, and C. W. Wessner. 2005. *Local Heroes in the Global Village: Globalization and the New Entrepreneurship Policies.* New York: Springer.

Augustine, C., et al. 2009. *Redefining What's Possible for Clean Energy by 2020.* Full Report. Gigaton Throwdown. June.

Bajaj, V. 2009. "India to spend $900 million on solar." *The New York Times* November 20.

Baldwin, J. R., and P. Hanel. 2003. *Innovation and Knowledge Creation in an Open Economy: Canadian Industry and International Implications.* Cambridge: Cambridge University Press.

Balzat, M., and A. Pyka. 2006. "Mapping national innovation systems in the OECD area." *International Journal of Technology and Globalisation* 2(1-2):158-176.

Bank of Boston. 1997. "MIT: The Impact of Innovation." Boston: Bank of Boston.

Battelle. 2009. *R&D Magazine*. December.

Battelle Technology Partnership Practice. 2009. "Opportunities for Advancing Job-Creating Research in Arkansas: A Strategic Assessment of Arkansas University and Government Lab Research Base." Access at <http://www.aralliance.org/__data/assets/pdf_file/0017/1682/Job-Creating-Research-in-Arkansas.pdf>.

Bezdek, R. H., and F. T. Sparrow. 1981. "Solar subsidies and economic efficiency." *Energy Policy* 9(4):289-300.

Biegelbauer, P. S., and S. Borras, eds. 2003. *Innovation Policies in Europe and the U.S.: The New Agenda.* Aldershot, UK: Ashgate.

Birch, D. 1981. "Who creates jobs?" *The Public Interest* 65:3-14.

Biris, Alexandru S. et al. 2009. "*In vivo* Raman flow cytometry for real-time detection of carbon nanotube kinetics in lymph, blood, and tissues." *J. Biomed Opt* 14(2).

Block, F., and M. R. Keller. 2008. *Where Do Innovations Come From? Transformations in the U.S. National Innovation System, 1970-2006.* Washington, DC: The Informational Technology & Innovation Foundation. July.

Blomström, M., A. Kokko, and F. Sjöholm. 2002. "Growth & Innovation Policies for a Knowledge Economy: Experiences from Finland, Sweden, & Singapore." EIJS Working Paper. Series No. 156.

Bloomberg News. 2006. "The next green revolution." August 21.

Bolinger, M., R. Wiser, and E. Ing. 2006. "Exploring the Economic Value of EPAct 2005's PV Tax Credits." Lawrence Berkeley National Laboratory.

Borenstein, S. 2008. *The Market Value and Cost of Solar Photovoltaic Electricity Production.* Berkeley, CA: Center for the Study of Energy Markets.

Borras, S. 2003. *The Innovation Policy of the European Union: From Government to Governance.* Cheltenham, UK: Edward Elgar.

Borrus, M., and J. Stowsky. 2000. "Technology policy and economic growth." In C. Edquist and M. McKelvey, eds. *Systems of Innovation: Growth, Competitiveness and Employment, Vol. 2.* Cheltenham, UK and Northampton, MA: Edward Elgar.

Bradsher, K. 2009. "China builds high wall to guard energy industry." *International Herald Tribune* July 13.

Brander, J. A., and B. J. Spencer. 1983. "International R&D rivalry and industrial strategy." *Review of Economic Studies* 50:707-722.

Brander, J. A., and B. J. Spencer. 1985. "Export strategies and international market share rivalry." *Journal of International Economics* 16:83-100.

Branigin, W. 2009. "Obama lays out clean-energy plans." *Washington Post* March 24. p. A05.

Branscomb, L., and P. Auerswald. 2002. *Between Invention and Innovation: An Analysis of Funding for Early-Stage Technology Development.* NIST GCR 02–841. Gaithersburg, MD: National Institute of Standards and Technology. November.

Braunerhjelm, P., and M. Feldman. 2006. *Cluster Genesis: Technology based Industrial Development.* Oxford: Oxford University Press.

Business Facilities. 2010. "Windstream Picks Little Rock, AR for HQ." July 13.

Bush, N. 2005. "Chinese competition policy, it takes more than a law." *China Business Review* May-June.

Bush, V. 1945. *Science: The Endless Frontier*. Washington, DC: Government Printing Office.

Bussey, J. 2012." The sun shines on 'the cloud.'" *The Wall Street Journal*. July 13:B1.

Campoccia, A., L. Dusonchet, E. Telaretti, and G. Zizzo. 2009. "Comparative analysis of different supporting measures for the production of electrical energy by solar PV and Wind systems: Four representative European cases." *Solar Energy* 83(3):287-297.

Caracostas, P., and U. Muldur. 2001. "The emergence of the new European Union research and innovation policy." In P. Laredo and P. Mustar, eds. *Research and Innovation Policies in the New Global Economy: An International Comparative Analysis*. Cheltenham, UK: Edward Elgar.

Carter, Mark. 2011. "Scholars Program Copies Georgia's Model." *Innovate Arkansas* August 22.

Chesbrough, H. 2003. *Open Innovation: The New Imperative for Creating and Profiting from Technology*. Cambridge, MA: Harvard Business School Press.

Chronicle of Higher Education. February 7, 2010.

Cimoli, M., and M. della Giusta. 2000. "The Nature of Technological Change and Its Main Implications on National and Local Systems of Innovation." IIASA Interim Report IR-98-029.

Clemson School of Computing. Undated. "Dr. Amy Apon Joins the School of Computing as Chair of the Computer Science Division." Press Release. Clemson, SC: Clemson University.

Cleveland.com. 2010. "Caterpillar Opens New Arkansas Factory, Hiring 600." September 1.

Coburn, C., and D. Berglund. 1995. *Partnerships: A Compendium of State and Federal Cooperative Programs.* Columbus, OH: Battelle Press.

Combs, K., and A. Link. 2003. "Innovation policy in search of an economic paradigm: the case of research partnerships in the United States." *Technology Analysis & Strategic Management* 15(2).

Computer Community Consortium. 2009. "From Internet to Robotics: A Roadmap for U.S. Robotics." May 21.

Constable, George, and Bob Somerville. 2003. *A Century of Innovatio: Twenty Engineering Achievements that Transformed our Lives.* Washington DC: Joseph Henry Press.

Cortright, J. 2006. *Making Sense of Clusters: Regional Competitiveness and Economic Development.* Washington, DC: Brookings Institution.

Cortright, J., and H. Mayer. 2002. *Signs of Life: The Growth of Biotechnology Centers in the US.* Washington, DC: Brookings Institution.

Council on Competitiveness/National Governor's Association. 2007. *Cluster-Based Strategies for Growing State Economies.* Washington, DC: Council on Competitiveness.

Crafts, N. F. R. 1995. "The golden age of economic growth in Western Europe, 1950-1973." *Economic History Review* 3:429-447.

Dahlman, C., and J. E. Aubert. 2001. *China and the Knowledge Economy: Seizing the 21st Century.* Washington, DC: World Bank.

Dahlman, C., and A. Utz. 2005. *India and the Knowledge Economy: Leveraging Strengths and Opportunities.* Washington, DC: World Bank.

Daniel, D. E. 2008. "Thoughts on Creating More Tier One Universities in Texas." White Paper. May 30.

Darmody, B. 2010. "The Power of Place 2.0: The Power of Innovation—10 Steps for Creating Jobs, Improving Technology Commercialization, and Building Communities of Innovation." Tucson: Association of University Research Parks. March 5.

Davis, S., J. Haltiwanger, and S. Schuh. 1993. "Small Business and Job Creation: Dissecting the Myth and Reassessing the Facts." Working Paper No. 4492. Cambridge, MA: National Bureau of Economic Research.

Debackere, K., and R. Veugelers. 2005. "The role of academic technology transfer organizations in improving industry science links." *Research Policy* 34(3):321-342.

Department of Labor and Industrial Relations: Research and Statistics Office. *Hawaii's Green Workforce: A Baseline Assessment.* December 2010.

Desai, S., P. Nijkamp, and R. R. Stough, eds. 2011. *New Directions in Regional Economic Development: The Role of Entrepreneurship Theory and Methods, Practice and Policy*. Northampton, MA: Edward Elgar.

De la Mothe, J., and G. Paquet. 1998. "National Innovation Systems, 'Real Economies' and Instituted Processes." *Small Business Economics* 11:101-111.

DeVol, R. C., K. Klowden, A. Bedorussian, and B. Yeo. 2009. *North America's High Tech Economy: The Geography of Knowledge-Based Institututions.* June 2.

De Vol, Ross et al. 2004. *Arkansas' Position in the Knowledge-Based Economy* Santa Monica: Milken Institute.

Dobesova, K., J. Apt, and L. Lave. 2005. "Are renewable portfolio standards cost-effective emissions abatement policy?" *Environmental Science and Technology* 39:8578-8583.

Doloreux, D. 2004. "Regional innovation systems in Canada: a comparative study." *Regional Studies* 38(5):479-492.

Doris, E., J. McLaren, V. Healey, and S. Hockett. 2009. *State of the States 2009: Renewable Energy Development and the Role of Policy.* Golden, CO: National Renewable Energy Laboratory.

Durham, C. A., B. G. Colby, and M. Longstreth. 1988. "The impact of state tax credits and energy prices on adoption of solar energy systems." *Land Economics* 64(4):347-355.

Eaton, J., E. Gutierrez, and S. Kortum. 1998. "European Technology Policy." NBER Working Paper 6827.

Economic Development Agency. 2011. "EDA Announces Registry to Connect Industry Clusters Across the U.S.; Harvard Business School Tool Designed to Assist Innovators and Small Business in Spurring Regional Economic Growth." October 6.

Edler, J., and S. Kuhlmann. 2005. "Towards one system? The European Research Area initiative, the integration of research systems and the changing leeway of national policies." *Technikfolgenabschätzung: Theorie und Praxis* 1(4):59-68.

Eickelpasch, A., and M. Fritsch. 2005. "Contests for cooperation: a new approach in German innovation policy." *Research Policy* 34:1269-1282.

Energy Information Administration. 2008. *Federal Financial Interventions and Subsidies in Energy Markets 2007*. Washington, DC: Energy Information Administration.

Energy Overviews. 2011. "Arkansas Wins $100 Million Wind Turbine Nacelle Plant." May 11.

Engardio, P. 2008. "Los Alamos and Sandia: R&D Treasures." *BusinessWeek*. February 11.

Engardio, P. 2009. "State Capitalism." *BusinessWeek*. February 6.

Etzowitz, H. 2008. *The Triple Helix: University-Industry-Government Innovation in Action.* London: Routledge.

European Commission. 2003. *Innovation in Candidate Countries: Strengthening Industrial Performance*. Brussels: European Commission. May.

Fangerberg, J. 2002. *Technology, Growth, and Competitiveness: Selected Essays*. Cheltenham, UK, and Northampton, MA: Edward Elgar.

Farrell, C., and M. Mandell. 1992. "Industrial Policy." *BusinessWeek*. April 4.

Featherstonhaugh, George William. 1844. *Excursion Through The Slave States, From Washington On The Potomac To The Frontier Of Mexico.* London: John Murray, Albemarle Street.

Federal Reserve of Chicago. 2007. "Can Higher Education Foster Economic Growth?—A Conference Summary. *Chicago Fed Letter*. March.

Feldman, M., and A. Link. 2001. "Innovation policy in the knowledge-based economy." In *Economics of Science, Technology and Innovation, Vol. 23*. Boston: Kluwer Academic Press.

Feldman, M., A. Link, and D. Siegel. 2002. *The Economics of Science and Technology: An Overview of Initiatives to Foster Innovation, Entrepreneurship, and Economic Growth*. Boston: Kluwer Academic Press.

Feser, E. 2005. "Industry Cluster Concepts in Innovation Policy: A Comparison of U.S. and Latin American Experience." *Interdiscliplinary Studies in Economics and Management, Vol. 4*. Vienna: Springer.

Fishback, B., C. A. Gulbranson, R. E. Litan, L. Mitchell, and M. Porzig. 2007. *Finding Business "Idols": A New Model to Accelerate Start-Ups*. Kauffman Foundation Report.

Flamm, K. 2003. "SEMATECH revisited: assessing consortium impacts on semiconductor industry R&D." In National Research Council. *Securing the Future: Regional and National Programs to Support the Semiconductor Industry*. Charles. W. Wessner, ed. Washington, DC: The National Academies Press.

Florida, R. 2002. *The Rise of the Creative Class.* New York: Basic Books.

Florida, R. 2004. "The World is Spiky." *Atlantic Monthly* October.

Fonfria, A., C. Diaz de la Guardia, and I. Alvarez. 2002. "The role of technology and competitiveness policies: a technology gap approach." *Journal of Interdisciplinary Economics* 13:223-241.

Foray, D., and P. Llerena. 1996. "Information structure and coordination in technology policy: a theoretical model and two case studies." *Journal of Evolutionary Economics* 6(2):157-173.

Fort Smith Times Record. 2010. "Mitsubishi Incentives Hit $83M." December 25.

Fox-Penner, Peter S., Marc Chupka, and Robert L. Earle. 2008. "Transforming America's Power Industry: The Investment Challenge 2010–2030." Brattle Group.

Freeman, C. 1987. *Theory of Innovation and Interactive Learning.* London: Pinter.

Friedman, T. 2005. *The World Is Flat: A Brief History of the 21st Century.* New York: W. H. Freeman.

Fry, G. R. H. 1986. "The economics of home solar water heating and the role of solar tax credits." *Land Economics* 62(2):134-144.

Fthenakis, V., J. E. Mason, and K. Zweibel. 2009. "The technical, geographical, and economic feasibility for solar energy to supply the energy needs of the US." *Energy Policy* 37(2):387-399.

Fullilove, M. T. 2005. *Root Shock: How Tearing Up City Neighborhoods Hurts America and What We Can Do About It.* New York: Ballantine Books.

Furman, J., M. Porter, and S. Stern. 2002. "The determinants of national innovative capacity." *Research Policy* 31:899-933.

Geiger, R. L., and C. M. Sá. 2009. *Tapping the Riches of Science: Universities and the Promise of Economic Growth.* Cambridge MA: Harvard University Press.

George, G., and G. Prabhu. 2003. "Developmental financial institutions as technology policy instruments: implications for innovation and entrepreneurship in emerging economies." *Research Policy* 32(1):89-108.

Grande, E. 2001. "The erosion of state capacity and European innovation policy: a comparison of German and EU information technology policies." *Research Policy* 30(6):905-921.

Grindley, P., D. Mowery, and B. Silverman. 1994. "SEMATECH and collaborative research: lessons in the design of high technology consortia." *Journal of Policy Analysis and Management* 13(4):723-758.

Grossman, G. M., and E. Helpman. 1994. "Endogenous innovation in the theory of growth." *The Journal of Economic Perspectives* 8(1):23-44.

Guidolin, M., and C. Mortarino. 2010. "Cross-country diffusion of photovoltaic systems: modelling choices and forecasts for national adoption patterns." *Technological Forecasting and Social Change* 77(2):279-296.

Gulbranson, C. A., and D. B. Audretsch. 2008. "Proof of Concept Centers: Accelerating the Commercialization of University Innovation." Ewing Marion Kauffman Foundation. January.

Hall, B. 2002. "The assessment: technology policy." *Oxford Review of Economic Policy* 18(1):1-9.

Hall, B. 2004. "University-Industry Research Partnerships in the United States." Kansai Symposium Report. February.

Hamilton, Gregory L., and Teresa A. McLendon. 2006. *Closing the Gap: An Examination and Analysis of Per Capita Personal Income in Arkansas.* August. Little Rock: University of Arkansas at Little Rock.

Harbour, K. 2011. "WV Biometrics: Fertile Ground for Innovation." Charleston, WV: West Virginia Department of Commerce.

Harris, William C. 2010. "Innovation lessons from Ireland." *Research-Technology Management* 53(1):35-39.

Hill, Edward et al. 2012. "Economic Shocks and Regional Economic Resilience." Pages 193-274 in M. Weir, N. Pindus, H. Wial and H. Wolman, eds. *Urban and Regional Policy and Its Effects, vol. 4: Building Resilient Regions.* Washington, DC: Brookings Institution Press.

Ho, Giang, and Anthony Pennington-Cross. 2005. "Fayetteville and Hot Springs Lead the Recovery in Employment." *The Regional Economist* October.

Hodges, Curt. 2011. "Beckmann Volmer Breaks Ground on Osceola Plant." *Paragould Daily Press* September 14.

Hu, Z. 2006. "IPR Policies in China: Challenges and Directions." Presentation at *Industrial Innovation in China*. Levin Institute Conference. July 24-26.

Hughes, K. 2005. *Building the Next American Century: The Past and Future of American Economic Competitiveness*. Washington, DC: Woodrow Wilson Center Press.

Hughes, K. 2005. "Facing the global competitiveness challenge." *Issues in Science and Technology* XXI(4):72-78.

Jaffe, A., J. Lerner, and S. Stern, eds. 2003. *Innovation Policy and the Economy, Vol. 3*. Cambridge, MA: MIT Press.

Janssen, M. A., R. Holahan, A. Lee, and E. Ostrom. 2010. "Lab Experiments for the Study of Social-Ecological Systems." *Science* 328(5978):613-617. April.

Jaruzelski, B., and K. Dehoff. 2008. "Beyond borders: The global innovation 1000." *Strategy and Business* 53(Winter).

Jasanoff, S., ed. 1997. *Comparative Science and Technology Policy*. Elgar Reference Collection. International Library of Comparative Public Policy, Vol. 5. Cheltenham, UK, and Lyme, NH: Edward Elgar.

Jorgenson, D., and K. Stiroh. 2002. "Raising the speed limit: economic growth in the information age." In National Research Council. 2002. *Measuring and Sustaining the New Economy*. Dale. W. Jorgenson and Charles. W. Wessner, eds. Washington, DC: The National Academies Press.

Joy, W. 2000. "Why the future does not need us." *Wired* 8(April).

Kelly, K. 1992. "Hot Spots." *BusinessWeek* October 19.

Kim, Y. 2006. "A Korean Perspective on China's Innovation System." Presentation at *Industrial Innovation in China*. Levin Institute Conference. July 24-26.

Koschatzky, K. 2003. "The regionalization of innovation policy: new options for regional change?" In G. Fuchs and P. Shapira, eds. *Rethinking Regional Innovation: Path Dependency or Regional Breakthrough?* London: Kluwer.

Krueger, A. O. "Globalization and International Locational Competition." Symposium in Honor of Herbert Giersch. Lecture delivered at the Keil Institute. May 11, 2006.

Kuhlmann, S., and J. Edler. 2003. "Scenarios of technology and innovation policies in Europe: investigating future governance—group of 3." *Technological Forecasting & Social Change* 70.

Lall, S. 2002. "Linking FDI and technology development for capacity building and strategic competitiveness." *Transnational Corporations* 11(3):39-88.

Lancaster, R. R., and M. J. Berndt. 1984. "Alternative energy development in the USA: the effectiveness of state government incentives." *Energy Policy* 12(2):170-179.

Laredo, P., and P. Mustar, eds. 2001. *Research and Innovation Policies in the New Global Economy: An International Perspective.* Cheltenham, UK: Edward Elgar.

Lee, Y. S. 2000. "The Sustainability of University-Industry Research Collaboration. *The Journal of Technology Transfer* 25(2).

Lerner, J. 1999. "Public venture capital." In National Research Council. *The Small Business Innovation Program: Challenges and Opportunities.* Charles. W. Wessner, ed. Washington, DC: National Academy Press.

Leslie, S. 1993. *The Cold War and American Science: The Military-Industrial-Academic Complex at MIT and Stanford*. New York: Columbia University Press.

Lewis, J. 2005. *Waiting for Sputnik: Basic Research and Strategic Competition.* Washington, DC: Center for Strategic and International Studies.

Lin, O. 1998. "Science and technology policy and its influence on the economic development of Taiwan." In H. S. Rowen, ed. *Behind East Asian Growth: The Political and Social Foundations of Prosperity*. London and New York: Routledge.

Link, A. N. 1995. *A Generosity of Spirit: The Early History of the Research Triangle Park.* Research Triangle Park: The Research Triangle Foundation of North Carolina.

Litan, R. E., L. Mitchell, and E. J. Reedy. 2007. "Commercializing University Innovations: Alternative Approaches." Boston: National Bureau of Economic Research. Working paper JEL No. O18, M13, 033, 034, 038.

Litan, R. E., L. Mitchell, and E. J. Reedy. 2007. "The University as Innovator: Bumps in the Road." *Issues in Science and Technology* Summer:57-66.

Lucas, R. "On the mechanics of economic development." *Journal of Military Economics* 22:38-39.

Luger, M. 2001. "Introduction: information technology and regional economic development." *Journal of Comparative Policy Analysis: Research & Practice.*

Luger, M., and H. A. Goldstein. 1991. *Technology in the Garden*. Chapel Hill: University of North Carolina Press.

Luger, M., and H. A. Goldstein. 2006. *Research Parks Redux: The Changing Landscape of the Garden*. Washington, DC: U.S. Department of Commerce.

Lundvall, B., ed. 1992. *National Innovation Systems: Towards a Theory of Innovation and Interactive Learning.* London: Pinter.

Luther, J. 2008. "Renewable Energy Development in Germany." Presentation at the NRC Christine Mirzayan Fellows Seminar. March 5, 2008. Washington, DC.

Maddison, A., and D. Johnston. 2001. *The World Economy: A Millennial Perspective*. Paris: Organization for Economic Co-operation and Development.

Mani, S. 2004. "Government, innovation and technology policy: an international comparative analysis." *International Journal of Technology and Globalization* 1(1).

Manufacturers' News. 2011. "Industrial Jobs in Arkansas Declined 1.5% Over Last Year." October 31.

Marshall, A. 1890. *Principles of Economics*. London: MacMillan & Company.

McKibben, W. 2003. *Enough: Staying Human in an Engineered Age*. New York: Henry Holt & Co.

McKinsey and Company. 2010 "Energy Efficiency, A Compelling Global Resource." McKinsey and Company.

Melissaratos, A., and N. J. Slabbert. 2009. *Innovation: The Key to Prosperity—Technology and America's Role in the 21st Century Global Economy*. Washington, DC: Montagu House.

Mendonca, M. 2007. *Feed-in Tariffs: Accelerating the Development of Renewable Energy*. London: Earthscan.

Meyer-Krahmer, F. 2001. "Industrial innovation and sustainability—conflicts and coherence." In D. Archibugi and B. Lundvall, eds. *The Globalizing Learning Economy*. New York: Oxford University Press.

Meyer-Krahmer, F. 2001. "The German innovation system." In P. Larédo and P. Mustar, eds. *Research and Innovation Policies in the New Global Economy: An International Comparative Analysis*. Cheltenham, UK: Edward Elgar.

Mills, K. G., E. B. Reynolds, and A. Reamer. 2008. *Clusters and Competitiveness: A New Federal Role for Stimulating Regional Economies*. Washington, DC: Brookings.

Moore, G. 2003. "The SEMATECH contribution." In National Research Council. *Securing the Future: Regional and National Programs to Support the Semiconductor Industry*. Charles. W. Wessner, ed. Washington, DC: The National Academies Press.

Moselle, B., J. Padilla, and R. Schmalensee. 2010. *Harnessing Renewable Energy in Electric Power Systems: Theory, Practice Policy*. Washington, DC: RFF Press.

Mufson, S. 2009. "Asian nations could outpace U.S. in developing clean energy." *Washington Post* July 16.

Murphy, L. M., and P. L. Edwards. 2003. *Bridging the Valley of Death: Transitioning from Public to Private Sector Financing*. Golden, CO: National Renewable Energy Laboratory. May.

Mustar, P., and P. Laredo. 2002. "Innovation and research policy in France (1980-2000) or the disappearance of the Colbertist state." *Research Policy* 31:55-72.

National Academy of Engineering. 2004. *The Engineer of 2020: Visions of Engineering in the New Century.* Washington, DC: The National Academies Press.

National Academy of Engineering. 2008. *Grand Challenges for Engineering.* Washington, DC: The National Academies Press.

National Academy of Sciences. 2010. *Electricity from Renewable Sources: Status, Prospects, and Impediments.* Washington, DC: The National Academies Press.

National Academy of Sciences, National Academy of Engineering, and Institute of Medicine. 2007. *Rising Above the Gathering Storm: Energizing and Employing America for a Brighter Economic Future.* Washington, DC: The National Academies Press.

National Academy of Sciences, National Academy of Engineering, and National Research Council. 2009. *America's Energy Future: Technology and Transformation.* Washington, DC: The National Academies Press.

National Academy of Sciences, National Academy of Engineering, and National Research Council. 2009. *Real Prospects for Energy Efficiency in the United States.* Washington, DC: The National Academies Press.

National Economic Council and Office of Science and Technology Policy. 2009. "A Strategy for American Innovation: Driving Towards Sustainable Growth and Quality Jobs." Washington, DC: Executive Office of the President. September.

National Governors' Association. 2007. *Innovation America.* Washington, DC: National Governors' Association.

National Institute of Standards and Technology. 2010. "NIST Manufacturing Extension Partnership Awards $9.1 Million for 22 Projects to Enhance U.S. Manufacturers' Global Competitiveness." Gaithersburg, MD: National Institute of Standards and Technology. October 5.

National Research Council. 1996. *Conflict and Cooperation in National Competition for High-Technology Industry.* Washington, DC: National Academy Press.

National Research Council. 1999. *The Advanced Technology Program: Challenges and Opportunities.* Charles. W. Wessner, ed. Washington, DC: National Academy Press.

National Research Council. 1999. *Funding a Revolution: Government Support for Computing Research.* Washington, DC: National Academy Press.

National Research Council. 1999. *Industry-Laboratory Partnerships: A Review of the Sandia Science and Technology Park Initiative.* Charles. W. Wessner, ed. Washington, DC: National Academy Press.

National Research Council. 1999. *New Vistas in Transatlantic Science and Technology Cooperation.* Charles. W. Wessner, ed. Washington, DC: National Academy Press.

National Research Council. 1999. *The Small Business Innovation Research Program: Challenges and Opportunities.* Charles. W. Wessner, ed. Washington, DC: National Academy Press.

National Research Council. 1999. *U.S. Industry in 2000: Studies in Competitive Performance*. D. C. Mowery, ed. Washington, DC: National Academy Press.
National Research Council. 2000. *The Small Business Innovation Research Program: A Review of the Department of Defense Fast Track Initiative.* Charles. W. Wessner, ed. Washington, DC: National Academy Press.
National Research Council. 2001. *A Review of the New Initiatives at the NASA Ames Research Center*. Charles. W. Wessner, ed. Washington, DC: National Academy Press.

National Research Council. 2001. *Building a Workforce for the Information Economy.* Washington, DC: National Academy Press.
National Research Council. 2001. *Capitalizing on New Needs and New Opportunities: Government-Industry Partnerships in Biotechnology and Information Technologies*. Charles. W. Wessner, ed. Washington, DC: National Academy Press.
National Research Council. 2001. *The Advanced Technology Program: Assessing Outcomes.* Charles. W. Wessner, ed. Washington, DC: National Academy Press.
National Research Council. 2001. *Trends in Federal Support of Research and Graduate Education.* S. A. Merrill, ed. Washington, DC: National Academy Press.
National Research Council. 2002. *Partnerships for Solid-State Lighting.* Charles. W. Wessner, ed. Washington, DC: National Academy Press.
National Research Council. 2003. *Government-Industry Partnerships for the Development of New Technologies: Summary Report.* Charles. W. Wessner, ed. Washington, DC: The National Academies Press.
National Research Council. 2003. *Securing the Future: Regional and National Programs to Support the Semiconductor Industry*. Charles. W. Wessner, ed. Washington, DC: The National Academies Press.
National Research Council. 2003. *Understanding Climate Change Feedbacks*. Washington, DC: The National Academies Press.
National Research Council. 2004. *Productivity and Cyclicality in Semiconductors: Trends, Implications, and Questions*. Dale W. Jorgenson and Charles. W. Wessner, eds. Washington, DC: The National Academies Press.
National Research Council. 2004. *The Small Business Innovation Research Program: Program Diversity and Assessment Challenges.* Charles. W. Wessner, ed. Washington, DC: The National Academies Press.
National Research Council. 2005. *Deconstructing the Computer*. Dale W. Jorgenson and Charles. W. Wessner, eds. Washington, DC: The National Academies Press.
National Research Council. 2005. *Getting Up to Speed: The Future of Superconducting*. S. L. Graham, M. Snir, and C. A. Patterson, eds. Washington, DC: The National Academies Press.

National Research Council. 2005. *Policy Implications of International Graduate Students and Postdoctoral Scholars in the United States*. Washington, DC: The National Academies Press.

National Research Council. 2006. *Software, Growth, and the Future of the U.S. Economy*. Dale W. Jorgenson and Charles. W. Wessner, eds. Washington, DC: The National Academies Press.

National Research Council. 2006. *The Telecommunications Challenge: Changing Technologies and Evolving Policies*. Charles. W. Wessner, ed. Washington, DC: The National Academies Press.

National Research Council. 2007. *Enhancing Productivity Growth in the Information Age: Measuring and Sustaining the New Economy*. Dale. W. Jorgenson and Charles. W. Wessner, eds. Washington, DC: The National Academies Press.

National Research Council. 2007. *Innovation Policies for the 21st Century*. Charles. W. Wessner, ed. Washington, DC: The National Academies Press.

National Research Council. 2007. *India's Changing Innovation System: Achievements, Challenges, and Opportunities for Cooperation*. Charles. W. Wessner and Sujai. J. Shivakumar, eds. Washington, DC: The National Academies Press.

National Research Council. 2007. *SBIR and the Phase III Challenge of Commercialization.* Charles. W. Wessner, ed. Washington, DC: The National Academies Press.

National Research Council. 2008. *An Assessment of the Small Business Innovation Research Program*. Charles. W. Wessner, ed. Washington, DC: The National Academies Press.

National Research Council. 2008. *An Assessment of the Small Business Innovation Research Program at the Department of Energy*. Charles. W. Wessner, ed. Washington, DC: The National Academies Press.

National Research Council. 2008. *An Assessment of the Small Business Innovation Research Program at the National Science Foundation*. Charles. W. Wessner, ed. Washington, DC: The National Academies Press.

National Research Council. 2008. *Innovative Flanders: Innovation Policies for the 21st Century*. Charles. W. Wessner, ed. Washington, DC: The National Academies Press.

National Research Council. 2008. *Innovation in Global Industries: U.S. Firms Competing in a New World*. J. Macher and D. Mowery, eds. Washington, DC: The National Academies Press.

National Research Council. 2008. *The National Academies Summit on America's Energy Future: Summary of a Meeting.* Washington, DC: The National Academies Press.

National Research Council. 2009. *21st Century Innovation Systems for Japan and the United States: Lessons from a Decade of Change*. S. Nagaoka, M. Kondo, K. Flamm, and C. Wessner, eds. Washington, DC: The National Academies Press.

National Research Council. 2009. *An Assessment of the Small Business Innovation Research Program at the Department of Defense*. Charles. W. Wessner, ed. Washington, DC: The National Academies Press.

National Research Council. 2009. *An Assessment of the Small Business Innovation Research Program at the National Aeronautics and Space Administration*. Charles. W. Wessner, ed. Washington, DC: The National Academies Press.

National Research Council. 2009. *An Assessment of the Small Business Innovation Research Program at the National Institutes of Health*. Charles. W. Wessner, ed. Washington, DC: The National Academies Press.

National Research Council. 2009. *Hidden Costs of Energy: Unpriced Consequences of Energy Production and Use*. Washington, DC: The National Academies Press.

National Research Council. 2009. *Revisiting the Department of Defense SBIR Fast Track Initiative*. Charles. W. Wessner, ed. Washington, DC: The National Academies Press.

National Research Council. 2009. *Understanding Research, Science and Technology Parks: Global Best Practices*. Charles. W. Wessner, ed. Washington, DC: The National Academies Press.

National Research Council. 2009. *Venture Funding and the NIH SBIR Program*. Charles. W. Wessner, ed. Washington, DC: The National Academies Press.

National Research Council. 2010. *Managing University Intellectual Property in the Public Interest*. Stephen Merrill and Anne-Marie Mazza, eds., Washington, DC: The National Academies Press.

National Research Council. 2011. *Building the 21st Century: U.S.-China Cooperation on Science, Technology, and Innovation*. Charles. W. Wessner, rapporteur. Washington, DC: The National Academies Press.

National Research Council. 2011. *Growing Innovation Clusters for American Prosperity*, Charles W. Wessner, rapporteur, Washington, DC: The National Academies Press.

National Research Council. 2011. *The Future of Photovoltaics Manufacturing in the United States.* Charles. W. Wessner, rapporteur. Washington, DC: The National Academies Press.

National Research Council. 2012. *Building Hawaii's Innovation Economy*. Charles. W. Wessner, rapporteur. Washington, DC: The National Academies Press.

National Research Council. 2012. *Clustering for 21st Century Prosperity*. Charles. W. Wessner, rapporteur. Washington, DC: The National Academies Press.

National Research Council. 2012. *Meeting Global Challenges: German-U.S. Innovation Policy*. Charles. W. Wessner, rapporteur. Washington, DC: The National Academies Press.

National Research Council. 2012. *Rising to the Challenge: U.S. Innovation Policy for the Global Economy*. Charles. W. Wessner and Alan Wm. Wolff, editors. Washington, DC: The National Academies Press.

National Research Council of Canada. 2008. *State of the Nation 2008: Canada's Science, Technology, and Innovation System*. Ottawa: Government of Canada.

National Science Board. 2010. *Science and Engineering Indicators 2010*. Arlington, VA: National Science Foundation.

Needham, J. 1954-1986. *Science and Civilization in China* (five volumes). Cambridge: Cambridge University Press.

Nelson, R., and K. Nelson. 2002. "Technology, institutions, and innovation systems." *Research Policy* 31:265-272.

Nelson, R., and N. Rosenberg. 1993. "Technical innovation and national systems." In R. R. Nelson, ed. *National Innovation Systems: A Comparative Analysis*. Oxford: Oxford University Press.

North American Windpower. 2010. "Mitsubishi Breaks Ground on Nacelle Facility in Arkansas." October 8.

NWA Online. 2010. "Firm Building Jonesboro Plant to Get $22 Million Stimulus." January 11.

O'Hara, M. P. 2005. *Cities of Knowledge: Cold War Science and the Search for the Next Silicon Valley*. Princeton: Princeton University Press.

O'Reilly, Joseph. 2009. "Arkansas: A Natural Wonder." *Inbound Logistics* May.

Organisation for Economic Co-operation and Development. 1997. "National Innovation Systems." Paris: Organisation for Economic Co-operation and Development.

Organisation for Economic Co-operation and Development. 2009. *Main Science and Technology Indicators*. Paris: Organisation for Economic Co-operation and Development.

Orszag, P., and T. Kane. 2003. "Funding Restrictions at Public Universities: Effects and Policy Implications." *Brookings Institution Working Paper*. September.

Oughton, C., M. Landabaso, and K. Morgan. 2002. "The regional innovation paradox: innovation policy and industrial policy." *Journal of Technology Transfer* 27(1).

Palmintera, D. 2005. *Accelerating Economic Development through University Technology Transfer*. Reston, VA: Innovation Associates.

Pavitt, K. 1998. "The Social Shaping of the National Science Base." *Research Policy* 27:793-805.

Pezzini, M. 2003. "Cultivating Regional Development: Main Trends and Policy Challenges in OECD Regions." Paris: Organisation for Economic Co-operation and Development.

Plastics News. 2011. "River Bend Gets Kosmo Work." November 7.

Porter, M. E. 1990. *The Competitive Advantage of Nations*. New York: The Free Press.

Porter, M. E., ed. 1993. *Choosing to Compete: A Statewide Strategy for Job Creation and Economic Growth*. Boston: The Commonwealth of Massachusetts.
Porter, M. E. 1998. "Clusters and the new economics of competition" *Harvard Business Review* 76(6):77-90.
Porter, M. E. 2005. *Clusters of Innovation Initiative: Regional Foundations of U.S. Competitiveness*. Washington DC: Council on Competitiveness.
Posen, A. 2001. "Japan." In B. Steil, D. G. Victor, and R. R. Nelson, eds. *Technological Innovation and Economic Performance*. Princeton: Princeton University Press.

President's Council of Advisors on Science and Technology. 2004. *Sustaining the Nation's Innovation System: Report on Information Technology Manufacturing and Competitiveness*. Washington, DC: Executive Office of the President. January.
PricewaterhouseCoopers. 2006. "China's Impact on the Semiconductor Industry: 2005 Update." PricewaterhouseCoopers.
PricewaterhouseCoopers and National Venture Capital Association. 2010. "MoneyTree Report." PricewaterhouseCoopers.
Pulaski County Chancery Court. 2001. *Lake View School District No. 25 v. Huckabee*. No. 1992-5318. May 25.
Purdue University. 2009. *Crossing the Next Regional Frontier: Information and Analytics Linking Regional Competitiveness to Investment in a Knowledge-Based Economy*. West Lafayette, IN: Purdue University. October.
Raduchel, W. 2006. "The end of stovepiping." In National Research Council. *The Telecommunications Challenge: Changing Technologies and Evolving Policies*, Charles. W. Wessner, ed., Washington, DC: The National Academies Press.
Ragwitz, M., and C. Huber. 2005. "Feed-in systems in Germany and Spain: a comparison." Fraunhofer Institut für Systemtechnik und Innovationsforschung.
Reid, T. R. 2004. *The United States of Europe: The New Superpower and the End of American Supremacy*. New York: Penguin Press.
Renewable Energy Policy Network for the 21st Century. 2009. *Renewables Global Status Report 2009*. Paris: REN21.
Rickerson, W., and R. Grace. 2007. "The Debate Over Fixed Price Incentives for Renewable Electricity in Europe and the United States: Fallout and Future Directions." White Paper prepared for the Heinrich Böll Foundation. Washington, DC.
Rolnick, Arthur J. 2007. "Congress Should End the Economic War Among the States." Testimony before the House Domestic Policy Subcommittee. October 10.
Romer, P. M. 1990. "Endogenous technological change." *Journal of Political Economy* October.

Rosenberg, N., and R. R. Nelson. 1994. "American universities and technical advance in industry." *Research Policy* 23:323-248

Ruttan, V. 2002. *Technology, Growth and Development: An Induced Innovation Perspective.* Oxford: Oxford University Press.

Rutten, R., and F. Boekema. 2005. "Innovation, policy and economic growth: theory and cases." *European Planning Studies* 13(8).

Sallet, J., E. Paisley, and J. R. Masterman. 2009. "The Geography of Innovation: the Federal Government and the Growth of Regional Innovation clusters." *Science Progress.* September.

Sarzynski, A. 2010. "The Impact of Solar Incentive Programs in Ten States." George Washington Institute of Public Policy Technical Report. Revised March 2010.

Saxenian, A. 1994. *Regional Advantage: Culture and Competition in Silicon Valley and Route 128.* Cambridge, MA: Harvard University Press.

Schumpeter, Joseph A. 1975. *Capitalism, Socialism, and Democracy.* New York: Harper.

Scott, A. J. 2004. *On Hollywood: The Place, the Industry.* Princeton NJ: Princeton University Press.

SERI (Solar Energy Industries Association). 2009. *U.S. Solar in Review 2008.* Washington, DC: Solar Energy Industries Association.

Shang, Y. 2006. "Innovation: New National Strategy of China." Presentation at Industrial Innovation in China. Levin Institute Conference. July 24-26.

Sheehan, J., and A. Wyckoff. 2003. "Targeting R&D: Economic and Policy Implications of Increasing R&D Spending." DSTI/DOC(2003)8. Paris: Organisation for Economic Co-operation and Development.

Sherwood, L. 2008. *U.S. Solar Market Trends 2007.* Latham, NY: Interstate Revewable Energy Council.

Small Business Administration. 2010. "SBA Announces Support for 10 Regional 'Innovative Economies' Clusters, Local Job Creation." SBA News Release 10-50. October 20.

Smits, R., and S. Kuhlmann. 2004. "The rise of systemic instruments in innovation policy." *International Journal of Foresight and Innovation Policy* 1(1/2).

Sparks, Glen R. 2007. "Community Profile: Conway, Ark. Makes Play for Economic Boom." *The Regional Economist* July.

Speck, S. 2008. "The design of carbon and broad-based energy taxes in European countries." *Vermont Journal of Environmental Law* 10.

Spencer, W., and T. E. Seidel. 2004. "International technology roadmaps: The U.S. semiconductor experience." In National Research Council. *Productivity and Cyclicality in Semiconductors: Trends, Implications, and Questions.* D. W. Jorgenson and Charles W. Wessner, eds. Washington, DC: The National Academies Press.

Stanford University. 1999. *Inventions, Patents and Licensing: Research Policy Handbook.* Document 5.1. July 15.

Stokes, D. E. 1997. *Pasteur's Quadrant: Basic Science and Technological Innovation.* Washington, DC: Brookings Institution.

Sturgeon, T. J. 2000. "How Silicon Valley Came to Be." In M. Kenney (ed.), *Understanding Silicon Valley: The Anatomy of an Entrepreneurial Region* (pp. 15-47). Stanford, CA: Stanford University Press.

Swamidass, P. M., and V. Vulasa. 2009. "Why university inventions rarely produce income? Bottlenecks in university technology transfer." *The Journal of Technology Transfer* 34(4).

Taleb, N. N. 2007. *The Black Swan: The Impact of the Highly Improbable.* New York: Random House.

Tan, J. 2006. "Growth of industry clusters and innovation: lessons from Beijing Zhongguancun Science Park." *Journal of Business Venturing* 21(6):827-850. November.

Task Force for the Creation of Knowledge-Based Jobs. 2002. *Report of the Task Force for the Creation of Knowledge-Based Jobs.* Accessed at <http://www.asta.arkansas.gov/resources/Documents/Knowledge-Based%20Jobs%20Report.pdf>. September.

Tassey, G. 2004. "Policy issues for R&D investment in a knowledge-based economy." *Journal of Technology Transfer* 29:153-185.

Taylor, M. 2008. "Beyond technology-push and demand-pull: lessons from California's solar policy." *Energy Economics* 30(6):2829-2854.

Teubal, M. 2002. "What is the systems perspective to innovation and technology policy and how can we apply it to developing and newly industrialized economies?" *Journal of Evolutionary Economics* 12(1-2).

Thomas, Kenneth P. 2011. *Investment Incentives and the Global Competition for Capital.* London and Basingstoke: Palgrave MacMillan.

Thompson, Susan C. 2010. "Factory Closing Shocks Community into Opening Wallets for Economic Development." *The Regional Economist.* October.

Tödtling, F., and M. Trippl. 2005 "One size fits all? Towards a differentiated regional innovation policy approach." *Research Policy* 34.

Tol, R. S. J. 2008. "The social cost of carbon: trends, outliers, and catastrophes." *Economics—the Open-Access, Open-Assessment E-Journal* 2(25):1-24.

Tzang, C. 2010. "Managing innovation for economic development in greater China: The origins of Hsinchu and Zhongguancun." *Technology in Society* 32(2):110-121. May.

University of Arkansas College of Engineering. 2008. "University of Arkansas Installing Supercomputer, 'Star of Arkansas', to be State's Fastest." Press Release. Fayetteville, AR: University of Arkansas at Fayetteville.

U.S. Department of Energy. 2006. Press Release. "Department Requests $4.1 Billion Investment as Part of the American Competitiveness Initiative: Funding to Support Basic Scientific Research." February 2.

U.S. General Accounting Office. 2002. *Export Controls: Rapid Advances in China's Semiconductor Industry Underscore need for Fundamental*

U.S. *Policy Review*. GAO-020620. Washington, DC: U.S. General Accounting Office. April.

Van Looy, B., K. Debackere, and T. Magerman. 2005. *Assessing Academic Patent Activity: The Case of Flanders.* Leuven: SOOS.

Van Looy, B., M. Ranga, J. Callaert, K. Debackere, and E. Zimmermann. 2004. "Combining Entrepreneurial and Scientific Performance in Academia: Towards a Compounded and Reciprocal Matthew-effect?" *Research Policy* 33(3):425-441.

Veugelers, R., J. Larosse, M. Cincera, D. Carchon, and R. Kalenga-Mpala. 2004. "R&D activities of the business sector in Flanders: results of the R&D surveys in the context of the 3% target." Brussels: IWT-Studies.

Wang, C. 2005. "IPR sails against current stream." *Caijing* October 17.

Wang, Q. 2010. "Effective policies for renewable energy—the example of China's wind power—lessons for China's photovoltaic power." *Renewable and Sustainable Energy Reviews* 14(2):702-712.

Wessner, C. W. 2005. "Entrepreneurship and the innovation ecosystem." In D. B. Audretsch, H. Grimm, and C. W. Wessner, eds. *Local Heroes in the Global Village: Globalization and the New Entrepreneurship Policies.* New York: Springer.

Wessner, C. W. 2005. *Partnering Against Terrorism.* Washington, DC: The National Academies Press.

Wind Power News. 2011. "A Wind Study the Size of Arkansas." April 1.

Wind Systems. 2011. Interview with Joe Brenner, Vice President of Nordex USA. January.

Wiser, R., G. Barbose, C. Peterman, and N. Darghouth. 2009. *Tracking the Sun II: The Installed Cost of Photovoltaics in the U.S. from 1998-2008.* Berkeley, CA: Lawrence Berkeley National Laboratory.

Witt, C. E., R. L. Mitchell, and G. D. Mooney. 1993. "Overview of the Photovoltaic Manufacturing Technology (PVMaT) Project." Paper presented at the 1993 National Health Transfer Conference. August 8-11. Atlanta, Georgia.

Xi, Lu Michael B. McElroy, and Juha Kiviluoma. 2009. "Global potential for wind-generated electricity." *Proceedings of the National Academy of Sciences of the United States of America* 106(27):10933-10938.

Yu, J., and R. Jackson. 2011. "Regional Innovation Clusters: A Critical Review." *Growth and Change* 42(2).

Zeigler, N. 1997. *Governing Ideas: Strategies for Innovation in France and Germany.* Ithaca, NY, and London: Cornell University Press.

Zweibel, K. 2010. "Should solar photovoltaics be deployed sooner because of long operating life at low, predictable cost?" *Energy Policy* 38(11):7519-7530.